A Throughput-Based Analysis of Army Active Component/ Reserve Component Mix for Major Contingency Surge Operations

Michael E. Linick, Igor Mikolic-Torreira, Katharina Ley Best, Alexander Stephenson, Jeremy M. Eckhause, Isaac Baruffi, Christopher Carson, Eric J. Duckworth, Melissa Bauman

Prepared for the Office of the Assistant Secretary of Defense for Reserve Affairs

For more information on this publication, visit www.rand.org/t/RR1516

Library of Congress Control Number: 2019939397

ISBN: 978-0-8330-9770-5

Published by the RAND Corporation, Santa Monica, Calif.

Preface

Because the needs of the United States and the threats it faces are continuously evolving, the balance between the active and reserve components of the U.S. armed forces is a perennial challenge for the U.S. Department of Defense. Aspects include how rapidly and in what quantity reserve component units can be made ready to deploy to meet the demands of a sudden, large overseas conflict, such as an Operation Desert Shield/Desert Storm–like event or a possible conflict on the Korean peninsula. The Office of the Secretary of Defense for Reserve Affairs asked RAND to explore these questions for Army units and to examine what can be done to maximize the number of ready forces from the Army's reserve components available to support such a conflict. This study focuses on the initial deployment speed of Army Reserve units, taking training requirements as an input, and assuming that trained and mobilized units will be ready to perform assigned roles once deployed.

The research in this report focuses on how the dynamics of the mobilization process for reserve component units can, or should, affect decisionmaking about force mix, as well as how policy and resourcing decisions can either enhance or inhibit the speed and efficiency of reserve component mobilization. It is intended to inform officials in the Office of the Secretary of Defense and the Department of the Army who are making decisions on the force structure balance between the active and reserve Army components or decisions on investments in readiness of the reserve components. The main body of this report includes technical details and assumes a familiarity with the concepts of mobilization planning and basic optimization. We aim to draw conclusions and insights in an accessible way throughout the report.

Our research is based on information gleaned from discussions with key Army commands and organizations involved in the planning and execution of force flows, mobilization, and training, combined with analytic modeling of the mobilization and postmobilization training processes. This work allowed us to identify those factors that have the greatest impact on increasing the ability to mobilize ready-to-deploy reserve component units in support of a major conflict, as well as those factors that have little effect on this.

This research was sponsored by the Office of the Assistant Secretary of Defense for Reserve Affairs. It was conducted within the Forces and Resources Policy Center of the RAND National Defense Research Institute, a federally funded research and development center sponsored by the Office of the Secretary of Defense, the Joint Staff, the Unified Combatant Commands, the Navy, the U.S. Marine Corps, the defense agencies, and the defense Intelligence Community.

For more information on the RAND Forces and Resources Policy Center, see http://www.rand.org/nsrd/ndri/centers/frp.html or contact the director (contact information is provided on the web page).

Contents

Figures

throughput of the RC training and mobilization pipeline have a greater effect on the number and types of RC units that can be deployed quickly.[4] These limitations on training and mobilization throughput over time are among several key considerations that could be used to inform decisions on AC/RC force mix.

To understand the effect that changes to policies and processes have on RC postmobilization throughput in the event of a large-scale, time-sensitive overseas contingency, RAND created two models of the mobilization and postmobilization training pipeline that could be used to simulate the effects that changing different factors would have on the pipeline. These models helped the research team identify the key drivers of postmobilization throughput, as well as major bottlenecks that prevent fast and efficient mobilization of the RC. We explore the effects of policy decisions related to planning, sequencing of units, and investments in premobilization readiness and postmobilization throughput capacity.

Findings

The research findings combine some intuitively obvious and some not-so-obvious insights about mobilization and deployment processes. First, we found that neither sealift capacity nor the demand signal is particularly problematic. TPFDD reasonably estimate the time available to mobilize ready RC units, and sufficient sealift capacity generally exists or can be made available to meet or come close to meeting deployment timelines specified in the TPFDD if optimistic but historically reasonable assumptions are made about commercial shipping access (see Appendix A). Instead, the major constraints on deploying RC units for a major time-sensitive contingency involve the mobilization process: (1) the time required for these units to complete their training after they are mobilized, which varies by unit and mission

[4] In this sense, capacity represents how many soldiers can be at the postmobilization stations simultaneously, and throughput represents how rapidly and in what quantity different types of RC forces can be made ready to deploy.

complexity, and (2) the physical capacity of the mobilization pipeline, which limits the number of units and personnel that can move through the mobilization process at one time.

Several actions can reduce delays caused by the mobilization process:

- mobilizing earlier if warning is received in advance and if early mobilization is politically and practically feasible
- opening and expanding mobilization capacity earlier
- keeping more training facilities in a state where they can ramp up quickly or increasing the speed with which capacity can be added
- shortening the postmobilization training time, either by changing the RC readiness posture (having units arrive at mobilization stations better trained and/or training them more efficiently while at the station); or having units deploy at lower readiness standards and accepting the risk
- focusing early mobilization on small, quicker-to-train units or on units that are well suited to rapid deployment.

However, some of these actions are more viable or less risky than others.

Earlier mobilization. The analysis shows that an early mobilization decision has by far the largest effect on the ability of RC units to mobilize, train, and deploy in time to meet the demands of a major war, but it is not always feasible.[5] Beginning mobilization eight weeks before the start of execution of the force flow can come close to doubling the number of both RC units and RC soldiers ready and able to meet the demands of a major war. The decision to mobilize is a political one that military planners cannot count on, but appropriate planning can allow senior leaders to take some steps to increase mobilization capacity and throughput prior to a mobilization decision. While

5 This observation should generally hold true in any case where the number of soldiers to be mobilized and processed through an MFGI exceeds the capacity of the facility, and therefore a set of sequencing and prioritization decisions are required. It is certainly true in major contingency surge operations that are the focus of this paper. For the rest of this report, *mobilization* refers to a large-scale mobilization for a major contingency surge operation.

not as effective as an early mobilization, these early actions to prepare the mobilization infrastructure will still have a positive effect on the speed of the overall mobilization process.

Time required for postmobilization training. This is the second biggest lever affecting the ability of RC units to meet the demands of a major war. However, its effect on the number of RC units ready and able to meet deployment demands is significantly less than that of early mobilization. Postmobilization training time is driven by several factors:

- postmobilization training requirements of the units involved, which are based on the unit's designed mission and complexity (e.g., large, complex units like brigade combat teams and combat aviation brigades that will conduct or support combined arms maneuver operations generally require more time than smaller, simpler units like the many combat support and combat service support units that deploy as companies or smaller elements)
- additional postmobilization training requirements particular to the specific mission or theater in which the unit will operate (e.g., are there additional mission-specific training requirements; must units be trained to C-1,[6] or is the risk of some lower readiness level acceptable?)
- readiness of incoming RC units (e.g., less-ready units will require more time than similar units entering the process in higher readiness states)[7]

---

[6] The Chairman's Readiness System defines several classifications of overall readiness, with C-1 as the highest, indicating that any issues or shortfalls the unit has will have negligible effect on its ability to execute assigned mission(s). Levels C-2, C-3, and C-4 indicate increasingly lower levels of readiness (see Chairman of the Joint Chiefs of Staff, *CJCS Guide to the Chairman's Readiness System*, Chairman of the Joint Chiefs of Staff Guide 3401D, Washington, D.C.: November 15, 2010). Army specific definitions of these levels are found in Army Regulation 220-1, *Army Unit Status Reporting and Force Registration—Consolidated Policies,* Washington, D.C.: Headquarters, Department of the Army, April 15, 2010.

[7] Readiness at mobilization can be a influenced by a wide variety of factors. We do not look at them discretely in this analysis, but note that the initial state of readiness is a critical variable. Among the key factors that do affect initial readiness is the question of personnel readiness—which encompasses the manning, military occupational specialty qualification,

- efficiency (or inefficiency) of postmobilization training processes (e.g., are the training sites able to conduct the needed training as planned or perhaps even accelerate it?).

Training site capacity. The analysis also shows that accelerating the rate at which training site capacity is expanded increases the production of RC units ready and able to meet deployment demands, but not in time to deliver significantly more ready RC units in the first five to six months of a conflict requiring a large mobilization of RC forces, especially one involving RC maneuver forces.

Risks of these approaches. Of course, each of these time-mitigating factors requires choices and may entail some risk. Optimistic assumptions about early mobilization (or expanding mobilization capacity) are difficult to defend. Most defense planning directs or assumes that the onset of mobilization (M-Day) will roughly coincide with an unambiguous warning (W-Day) or with the onset of combat operations (C-Day).[8] Assuming an even earlier mobilization decision will happen, it is possible that the share of RC forces contributing to the operation could be greater, and this could in turn have AC/RC force mix implications. However, this is a high-risk assumption; if much of the force structure is in the RC when a crisis occurs and early mobilization is not authorized, then there is little ability to get the required forces deployed in time. The other levers simply are not enough to compensate, even if money is unlimited once the war starts. This risk can be mitigated somewhat (and the decision to mobilize early potentially made more acceptable) by identifying those RC units that are needed to support the TPFDD and selectively mobilizing those units and the

---

and deployability of soldiers in the unit. Interviews with mobilization practitioners who had had experience with the initial surge or RC mobilizations in 2003 indicated that these three issues were the most important factors affecting speed of postmobilization training and validation.

[8] For a depiction of the notional mobilization process used to inform senior leadership, see Louis G. Yuengert, *How the Army Runs: A Senior Leader Reference Handbook 2015–2016*, Carlisle, Pa.: U.S. Army War College, 2015, pp. 5–8). This diagram shows partial mobilization taking place after the start of a conflict, with full-mobilization occurring well into the war.

training sites. However, this requires sufficiently early warning; significant planning; and a level of coordination across Army components, geographic COCOMs, and TRANSCOM that does not exist today.

Changing the readiness profile of the RC may require a higher level of resources and greater time commitment on the part of the RC soldiers. It is not clear (and not within the purview of this analysis) whether the RC could achieve or sustain the levels of premobilization readiness across a wide portion of the force necessary to significantly speed the throughput of the Army's mobilization stations. And, of course, deploying at lower levels of readiness to achieve the desired velocity and throughput may also incur significant risk, depending on where on the battlefield those units will deploy and the tasks and environment that will characterize their employment. Influencing some of these factors may require long lead-time investments (e.g., raising the overall readiness of RC units). Other factors might be managed in real time (accepting the risk of training units to lower readiness). But, collectively, they provide the Department of Defense with a variety of means to influence the availability of RC forces in support of a major conflict.

Which units go first. Additionally, how force managers envision sequencing units through the mobilization process also represents a set of choices. It is clear from our analysis that, given today's relatively low-level and slow ramp-up of mobilization throughput capacity, the mix and sequencing of RC units put through the mobilization pipeline has a significant effect on how many ready RC units can be produced in time. Inserting large units, such as RC brigade combat teams (BCTs) and combat aviation brigades (CABs), into the mobilization pipeline early in the process precludes smaller RC units from being delivered on time because RC BCTs and CABs both consume essentially all of the mobilization and postmobilization training capacity that the Army currently plans to have available early in a contingency, while also taking a relatively long time to train. Furthermore, because BCTs and CABs take so long to train, very few emerge from the mobilization pipeline in time to meet TPFDD demands, even if they are the first units mobilized. This means that the larger and more complex the unit desired, the fewer units could be accommodated and the slower the delivery. Small, relatively quicker-to-train units, in contrast, can move through the mobilization system fairly quickly.

Changing the AC/RC structure. The implications of these factors are less clear. They imply a somewhat different calculus for allocating RC units to the TPFDD than is used for the AC. This, in turn, has implications for both the structure of the RC and its readiness profile. The analysis in this report—specifically, the finding that only a few (if any) RC BCTs and CABs can be mobilized and made ready in time to meet TPFDD demands, regardless of when the mobilization decision is made—suggests there should be some bias in AC force structure toward BCTs and CABs and some bias in the RC toward smaller units that can be trained more quickly. Historical deployment data from recent large contingency operations suggest that the Army already relies on RC enabler units over RC maneuver units.[9] However, even smaller, quicker-to-train units may not be able to respond in the earliest days of a force flow. Of course, the large, short-warning contingency operation represents only one of many mission areas in which the RC can contribute. There are other considerations that influence the mix across the Total Force (such as homeland defense, stability operations, or rotational support to a long war, to name just a few) that argue for maintaining a number of BCTs and CABs in the RC. Nevertheless, the analysis provides some additional understanding about RC usage and Total Force structure.

Finally, it is important to note that, with *no* early mobilization, smaller RC units (with some, but limited, exceptions) will have trouble meeting demands in the first 30 to 60 days of TPFDD execution. This suggests that the set of enablers for deploying AC forces that are needed in the initial 30 to 60 days should either be maintained in the AC or kept in the RC but maintained at especially high readiness and be prepared to deploy on short notice.

---

[9] See Gregory Fontenot, E.J. Degen, and David Tohn, *On Point: The United States Army in Operation Iraqi Freedom*, Fort Leavenworth, Kan.: Combat Studies Institute Press, 2004, pp. 72–73.

Recommendations

Based on the findings of our analysis, we recommend that the Army planners and the global force managers responsible for allocating specific units to a major conflict contingency force flow use a process that recognizes the advantages of deploying smaller, quicker-to-train RC units in the earlier periods of a conflict and deferring the use of the larger, more-complex-to-train RC formations to later stages of major operations or transition or stabilization operations. We also recommend force managers focus investments on maintaining readiness in RC units based on the force flow imperatives suggested earlier. These investments may include not only training dollars but also training seat allocations, overmanning, and other actions that improve the general readiness of units when confronted with a no-notice or short-notice mobilization. While increases in postmobilization training capacity, as well as the ability to ramp up such capacity quickly, could also improve RC postmobilization training throughput, our analysis suggests that the small increases that would be reasonable (given the existing capacity and level of investment) would have a much lesser effect than the type of sequencing and readiness decisions mentioned earlier. Even a large investment in maintaining a much greater margin of contingency capacity for immediate mobilization does not significantly improve a mobilization process in which training time is the dominant factor. Finally, we recommend the Department of Defense consider reinstituting processes that match specific units (at the unit identification code level) to the operationally driven TPFDD demands to better focus peacetime and postmobilization training.

Caveats

There are some caveats to our findings and recommendations. Cost is not a factor we explicitly considered, although distinct cost differences can be inferred in the different investment options we profile. Also, this analysis assumes C-1 is the standard for all RC unit deployments, but the analysis explores the effect of significantly shortening training times; one way to achieve this would be training to a lower standard (such as C-2). The reduced training time increases the throughput of RC units, but whether the risk associated with lower training standards is acceptable is beyond the scope of this study. However, this caveat seems to reinforce the findings: The types of units that are able to deploy on time at less than C-1 (assuming that risk is acceptable) generally tend to be the smaller support units, not the larger combat units. This analysis is focused on the Army's RCs, and results are not necessarily generalizable to other services. Finally, and perhaps most importantly, this analysis is about contingency response for overseas wartime/conflict demands. It does not account for domestic missions (either defense support of civil authorities or homeland defense), such nonconflict deployments as humanitarian assistance or disaster relief, or "known" deployments (global force management sourced and approved). Each of these represents ripe ground for RC usage that may also affect AC/RC force mix decisionmaking.[10]

[10] For a more detailed look at some of the issues affecting AC/RC force mix decisions, see Chris Marie Briand, *Operational Army Reserve Sustainability Fact or Fiction?*, Carlisle, Pa.: U.S. Army War College, 2016.

Acknowledgments

Numerous individuals inside and outside the U.S. Department of Defense provided valuable assistance to our work. First, we thank Robert Smiley, Office of the Assistant Secretary of Defense for Reserve Affairs, for sponsoring this analysis. We also thank COL Robert Moore for guiding and supporting our research effort.

We are especially grateful for the assistance provided by leaders and subject-matter experts from the many entities critical to reserve component mobilization, which include Headquarters, Department of the Army, G-3/5/7 Force Provider Division; Army National Guard G-3 Mobilization and Readiness Division; U.S. Army Forces Command G-3/5/7 Force Provider Division; Directorate of Plans, Training, Mobilization and Security, Camp Shelby Joint Forces Training Center; 29th Infantry Division Headquarters, Virginia Army National Guard; Headquarters, First Army, Deputy Chief of Staff G-3/5/7; Directorate of Plans, Training, Mobilization and Security, U.S. Army Garrison Fort Hood; and Headquarters, First Army Division West G-3 Mobilization. All were invaluable in providing the data, personal perspectives, and feedback necessary for refining, verifying, and validating the mobilization throughput models used in our analysis.

Finally, at RAND, we thank Dan Madden, Jose Rodriguez, Zackary Steinborn, and Joshua Klimas for their contributions. They enhanced our analysis by providing training times and TPFDD cleaning for building supply and demand signals, overall model feedback, and administrative support, without which this effort would not have been possible. This paper has benefited greatly from the insights and recommendations provided by our reviewers, Josh Klimas,

Ed Filiberti, John Boon, Laura Baldwin, and Craig Bond. Finally, Maria Vega's gifted editorial touch helped hone the clarity of our presentation of the analysis.

CHAPTER ONE

Introduction

RAND was tasked by the Office of the Assistant Secretary of Defense for Reserve Affairs (OASD[RA]) with exploring the question of how well the processes for mobilizing and readying forces from the reserve components (RCs) align with timelines for when forces are needed on the ground in a future major conflict overseas. In particular, OASD(RA) asked RAND to look at the factors that would affect the active component (AC) and RC mix of Army forces deployed in response to a major contingency, in light of the time and resource requirements of both the strategic lift and mobilization processes. OASD(RA)'s question focuses on a future major overseas conflict requiring the rapid deployment of substantial combat, support, and sustainment forces with little or no notice. Examples would be an Operation Desert Storm–like event or a possible conflict on the Korean Peninsula.

To address this question, RAND developed an analytic methodology connecting future demands for capabilities on the ground in major contingencies, the transportation-constrained ability to deliver those capabilities into a theater of operations, and the ability to train and prepare the units providing those capabilities within the required timeline.[1] The fundamental tension that RAND analyzed is between

[1] Throughout this report, the term *training* refers to the activities between initial alert or mobilization and entering the warfight on the ground. This loose definition of *training* includes such activities as mobilization at home station, movement to mobilization station, soldier readiness processing, post-training validation, and other activities beyond unit-focused training. AC units also may go through a predeployment training process—absent the need to conduct mobilization activities or to move from home station to a training site.

the time when forces need to be ready to meet the demands of the conflict and the time required to mobilize and ready RC units to meet those demands. Setting overall force structure and overall readiness questions aside, the key factors that drive deployment speed are the availability of transportation and shipping capacity (known as *strategic lift*) and the time it takes for mobilization and training to be conducted. Of course, these two are interrelated: If limited strategic lift delays the movement of large combat units from the United States to the conflict location, there would be additional time to mobilize and train RC units, possibly making more of them available for the conflict; conversely, if there is an abundance of lift, it is possible that those platforms would be left waiting for units to complete predeployment training and arrive at the port of departure.

Analyses of existing planning scenarios and force flows reveal that transportation and shipping constraints factor into the Combatant Commanders' planning processes and the generation of time-phased deployment requirements, so the force flow requirements established by major war plans are all a transportation-feasible and mission-feasible compromise between what the Combatant Commander would like to do and what the transportation system can support.[2] This transportation-feasible flow may require the use of commercial shipping and other emergency capacity (see Appendix A). Given such extra capacity, planners expect that there is no "extra" time to prepare deploying units beyond what is specified in the transportation-feasible deployment plans; any failure to train and ready units in time for the transportation-feasible deployment plan would act as a bottleneck for deployment speed.

---

[2] For older discussions on transportation feasibility, see Chairman of the Joint Chiefs, *Joint Operation Planning and Execution System (JOPES), Vol. 1, (Planning Policies and Procedures)*, Chairman of the Joint Chiefs of Staff Manual 3122.01, Washington, D.C.: July 14, 2000, Change 1, May 25, 2001. For more recent policy and discussion on transportation risk informed planning, see Chairman of the Joint Chiefs of Staff, *Adaptive Planning and Execution Overview And Policy Framework*, Chairman of the Joint Chiefs of Staff Guide 3130, Washington, D.C.: May 29, 2015; and Chairman of the Joint Chiefs, *Campaign Planning Procedures and Responsibilities*, Chairman of the Joint Chiefs of Staff Staff Manual 3130.01A, Washington, D.C.: November 25, 2014.

For reasons further explained later in this chapter, the questions examined in this report are primarily of interest to the Army, which comprises the AC and two RCs, the Army National Guard and the Army Reserve. We have focused the analysis exclusively on the Army's two RCs and how they differ from the Army AC. Because relevant mobilization and training timelines are similar across the two RCs, we do not analyze National Guard and Army Reserve units separately.

The Army's ability to meet transportation-feasible force flow requirements for a large contingency operation depends on existing readiness and the ability to train and mobilize quickly. Predeployment training and preparation requirements generally apply to AC and RC units alike. That said, the requirements pose a more significant constraint to using RC units on short notice, because RC units must generally complete a larger portion of their training and preparation after they mobilize. This limits how rapidly they can deploy compared with AC units that have completed training. Moreover, the time required to get ready generally differs based on unit type and the complexity of the mission, among other factors. This means that smaller support and sustainment units generally take less time to get ready than larger units with combat missions.

AC and RC units are deployed using two largely separate training pipelines; AC and RC units are made ready for deployment independently and generally do not compete for the same housing, sustainment, soldier support, or trainers. The timing of RC mobilization is primarily determined by the demand signal, with mobilization dates linked to the latest arrival date (LAD) established by the Combatant Commander. However, strict adherence to a sequence of LADs may not be the most effective way to establish the ordering of units to the mobilization stations. The capacity and throughput of the RC mobilization and training pipeline have a greater effect on the number and types of RC units that can be deployed quickly. The limitations of mobilization and training throughput over time are among several key considerations informing decisions on the AC/RC force mix.

Background

> It must be clearly understood that implicit in the Total Force Policy, as emphasized by the Presidential and National Security Council documents, the Congress and the Secretary of Defense policy, is the fact that the Guard and Reserve forces will be used as the initial and primary augmentation of the Active forces.[3]
>
> —Total Force Policy memo by Secretary of Defense James Schlesinger

In issuing the Total Force Policy memo in 1973, Schlesinger noted, "Total Force is no longer a 'concept.' It is now the Total Force Policy which integrates the Active, Guard and Reserve forces into a homogenous whole."[4] The secretary also highlighted the need to meet readiness and deployment response times and asked for a study covering issues that shape the Total Force, including "availability, force mix, limitations and potential of Guard and Reserve Forces."[5]

These factors shaping the Total Force have been revisited repeatedly in the years since. In the early 1990s, Congress required the Secretary of Defense to provide "an assessment of a wide range of alternatives relating to the structure and mix of active and reserve forces appropriate for carrying out assigned missions in the mid- to late-1990s."[6] The resulting report, coordinated and led by RAND's Bernie Rostker, highlighted four characteristics for analysis of the problem: (1) the purpose of the military force, (2) the national military strategy in place, (3) the criteria for structuring forces (especially focusing on

---

[3] James Schlesinger, "Readiness of the Selected Reserves," Secretary of Defense memorandum, U.S. Department of Defense, August 23, 1973. (Underline emphasis appeared in original memo.)

[4] Schlesinger, 1973.

[5] Schlesinger, 1973.

[6] Public Law 102-90, National Defense Authorization Act for Fiscal Years 1992–1993, Title IV, Part A, Section 402, Assessment of the Structure and Mix of Active and Reserve Forces, 102nd Congress, December 5, 1991.

cost-effectiveness and political criteria), and (4) how active and reserve forces are integrated.[7] A 1995 study looked at "the adequacy of the evolving Army force structure, both active and reserve, to meet timetables for preparing combat and support forces to meet requirements" of the major regional contingencies; it concluded that the mix of combat and support forces in the RC should be re-examined, as should the way in which RC readiness was sustained.[8]

Since the mid-1990s—and especially since the onset of the global war on terrorism—the RC has transitioned from "strategic" forces that were called up once in a generation to "operational" forces that deploy far more frequently, and this transition is generating even more inquiries into how to envision and fund the Total Force. A Commission on the National Guard and Reserves, called for in the fiscal year 2005 National Defense Authorization Act,[9] established seven criteria for evaluating changes to law and policy with regard to the National Guard. Two of those criteria included answering whether the reforms (1) "improve the ability of the National Guard to meet both its overseas and homeland responsibilities as required by U.S. military, defense, and homeland security strategies" and (2) "enhance the ability of the National Guard to be a reserve force overseas, backing up the active

---

[7] Bernard Rostker, Charles Robert Roll, Marney Peet, Marygail Brauner, Harry J. Thie, Roger Allen Brown, Glenn A. Gotz, Steve Drezner, Bruce W. Don, Ken Watman, Michael G. Shanley, Fred L. Frostic, Colin O. Halvorson, Norman T. O'Meara, Jeanne M. Jarvaise, Robert Howe, David A. Shlapak, William Schwabe, Adele Palmer, James H. Bigelow, Joseph G. Bolten, Deena Dizengoff, Jennifer H. Kawata, Hugh G. Massey, Robert Petruschell, Craig Moore, Thomas F. Lippiatt, Ronald E. Sortor, J. Michael Polich, David W. Grissmer, Sheila Nataraj Kirby, and Richard Buddin, *Assessing the Structure and Mix of Future Active and Reserve Forces: Final Report to the Secretary of Defense*, Santa Monica, Calif.: RAND Corporation, MR-140-1-OSD, 1992.

[8] Ronald E. Sortor, *Army Active/Reserve Mix: Force Planning for Major Regional Contingencies*, Santa Monica, Calif.: RAND Corporation, MR-545-A, 1995, p. xii.

[9] Public Law 108-375, Ronald W. Reagan National Defense Authorization Act for Fiscal Year 2005, Title V, Subtitle B, Section 513, Commission on the National Guard and Reserves, 108th Congress, October 28, 2004.

force, and an operational force in the homeland, backed up by the active force."[10] In 2014, the Congressional Research Service noted:

> Determining the appropriate mix of AC and RC forces is complex, with many factors affecting the process. Of these, utilization, readiness, effectiveness, cost, and risk are generally considered the major elements in developing the AC/RC force mix.[11]

This set of criteria was similar to that noted by the Offices of the Secretary of Defense (OSD) in a 2013 report to Congress that said,

> There are several important factors in [AC] and [RC] mix decisions, including the timing, duration, and skills required for anticipated missions. Cost is always considered but is only one factor among many.[12]

Several other recent studies have looked at the relative costs of AC and RC force structure, including factoring in the relative availability of the forces.[13]

One key assumption in many of these studies has been that the mobilization capacity and transportation capacity existed to deploy RC forces "as needed". Several other studies have tested that assumption.[14]

---

[10] Commission on the National Guard and Reserves, "Strengthening America's Defense in the New Security Environment: Second Report to Congress," March 1, 2007, pp. 4–5.

[11] Andrew Feickert and Lawrence Kapp, "Army Active Component (AC)/Reserve Component (RC) Force Mix: Considerations and Options for Congress, Washington D.C.: *Congressional Research Service*, December 5, 2014.

[12] Chuck Hagel, "Unit Cost and Readiness For the Active and Reserve Components of the Armed Forces: Report to the Congress," Washington D.C., December 20, 2013, p. 3.

[13] See Joshua Klimas, Richard E. Darilek, Caroline Baxter, James Dryden, Thomas F. Lippiatt, Laurie L. McDonald, J. Michael Polich, Jerry M. Sollinger, and Stephen Watts, *Assessing the Army's Active-Reserve Component Force Mix*, Santa Monica, Calif.: RAND Corporation, RR-417-1-A, 2014; U.S. Government Accountability Office (GAO), *Active and Reserve Unit Costs,* GAO-14-711R, Washington, D.C., July 31, 2014; Hagel, 2013.

[14] See, for example, Ellen M. Pint, Matthew W. Lewis, Thomas F. Lippiatt, Philip Hall-Partyka, Jonathan P. Wong, and Tony Puharic, *Active Component Responsibility in Reserve Component Pre- and Postmobilization Training,* Santa Monica, Calif.: RAND Corporation, RR-738-A, 2015; Thomas F. Lippiatt and J. Michael Polich, *Reserve Component Unit*

However, in most cases, these studies tend to focus on how to best allocate training time and resources between premobilization and post-mobilization training. In other words, the issue is how to minimize the time or maximize the value of postmobilization training so that RC units can be "more responsive" to difficult deployment timelines. When resource constraints are identified, they tend to be in terms of trainers and training ranges, not overall mobilization capacity. Overall capacity is limited not only by trainers and training ranges, but also by the time it takes to open the mobilization facility (food, water, technology infrastructure, bed space, and other factors), the time it takes to arrive at the mobilization facility, completion of paperwork and verification of readiness, and the availability of such support staff as medical professionals and cooks. A salient exception to this postmobilization training time focus was a 2014 study conducted by the Surface Deployment and Distribution Command Transportation Engineering Agency (SDDC-TEA). The 2014 report contained some significant findings about shortfalls in planning and resourcing for mobilization capacity and its usage to support contingency operations. It also noted issues with RC readiness and how it aligned with projected mobilization processes and capacity.[15]

Research Focus

The research in this report focuses on how the capacity to mobilize and deploy RC forces can, or should, affect decisionmaking about investments in the Total Force and the potential options available to decisionmakers. The research examines overall mobilization capacity

---

Stability: Effects on Deployability and Training, Santa Monica, Calif.: RAND Corporation, MG-954-OSD, 2010; GAO, *Reserve Forces: Army Needs to Reevaluate its Approach to Training and Mobilizing Reserve Component Forces,* Washington, D.C.: GAO, GAO-09-720, 2009; Thomas F. Lippiatt, J. Michael Polich, and Ronald E. Sortor, *Post-Mobilization and Training of Army Reserve Component Combat Units,* Santa Monica, Calif.: RAND Corporation, MR-124-A, 1992.

[15] SDDC-TEA, *Reserve Component Mobilization Process and Requirements for Installation Infrastructure,* April 2014.

timing and resource constraints beyond trainers and training facilities. Specifically, the research looks to understand how several factors interact, including RC readiness and decisions about premobilization and postmobilization training; changes in training requirements and any associated risk; mobilization capacity, its elasticity, and the speed with which capacity can be increased; and mobilization policies, such as the timing of mobilization decisions and the sequencing of units for mobilization. The research focused on relatively large-scale deployments and used TPFDD derived from OSD-approved scenarios and concept plans (CONPLANs) to help test the models created to analyze this problem set.

While this research addresses, in principle, a broad question applying to all of the services, in practice the question is primarily of interest to the Army active, reserve, and guard components for the following reasons.

- Only the Army and Air Force have significant RCs that may "compete" with ACs to fill demands in theater. The Navy has a very limited RC, and the Marine Corps employs its RC primarily to "round out" active units.
- The Air Force Reserve and the Air National Guard have a different readiness and deployment model. They are less constrained overall by the factors described here and are difficult to divorce from the AC because individual aircraft and crews from the reserve components are integrated into active units in wartime deployment plans. While there are active debates as to the precise balance and mix of capabilities between the ACs and RCs, the ability of the Air Force RC to meet deployment requirements is generally accepted and not a major factor in these debates.[16]
- Discussions of Army force structure mix across the active, reserve, and National Guard components often include as a key issue the

[16] See, for example, the National Commission on the Structure of the Air Force's conclusion that the Air Force has made sufficient investments in individual- and crew-level RC readiness to allow this sort of integration (see National Commission on the Structure of the Air Force, *Report to the President and Congress of the United States*, Washington, D.C., January 30, 2014).

ability of reserve and guard elements to mobilize forces quickly enough to meet wartime demands.

For these reasons, this analysis is focused exclusively on the Army Reserve and the Army National Guard and will generally refer to these collectively as the *reserve components*. Because relevant mobilization and training timelines are similar across the two RCs, we do not analyze National Guard and Army Reserve units separately. This is not to imply that the two organizations are the same, but rather to emphasize that the fundamental subject of this study is the difference between the active Army component and the two reserve organizations, collectively.

Organization of This Report

This report is organized in four chapters and three appendixes. This first chapter introduces the problem and presents relevant background and context. Chapter Two describes the overall research methodology, explaining the two-step analytic approach used to first identify the key factors affecting mobilization throughput and then investigate those factors in more detail. Chapter Three describes the two models developed to support the analysis: the first a low-fidelity model used to identify the most important factors, and the second a high-fidelity model designed around those factors. The insights obtained from both models provide the basis for the findings and recommendations, which are presented in Chapter Four. Supporting information and analyses that are not central to the argument are presented in the three appendixes. Appendix A explains why postmobilization training and preparation issues, not shipping constraints, are the primary limiting factors for RC deployment speed. Appendix B provides the mathematical details of the high-fidelity model. Finally, Appendix C shows that different assumed readiness states of AC forces do not affect our findings and recommendations.

CHAPTER TWO

Analysis Methodology

In this chapter, we give a conceptual overview of the methodology used for this study. We introduce the conceptual model of the RC mobilization and training process that underlies all modeling work, and then give a brief introduction of the two models developed for this study. The low-fidelity model is a simulation tool that allowed us to explore the effect of broad policy levers on RC postmobilization training throughput. A high-fidelity model was then developed to fill in analytic gaps identified by low-fidelity model runs. A more detailed discussion of the modeling inputs, model structure, and outputs from the two models are presented in Chapter Three.

Our research began by developing a conceptual model of the overall training and mobilization processes used for delivering capabilities to meet a set of demands. Figure 2.1 shows one way of envisioning the force generation process.

It begins in the upper left with a defined time-phased demand signal, which identifies both the capabilities needed (units by type) and the timeline within which they must be delivered. Typically, this demand signal is described as the TPFDD—time-phased force and deployment data. The "supply" available to meet this demand signal can potentially be satisfied with either AC forces (middle column) or RC forces (right column). However, not all forces will be ready to deploy at the start of an operation. Some AC units will have completed most or all necessary training and will be able to deploy immediately or in short order; other AC units may need to complete training to be ready to deploy later in the force flow. On the RC side, even the

Figure 2.1
Conceptual Model of Training and Mobilization Processes

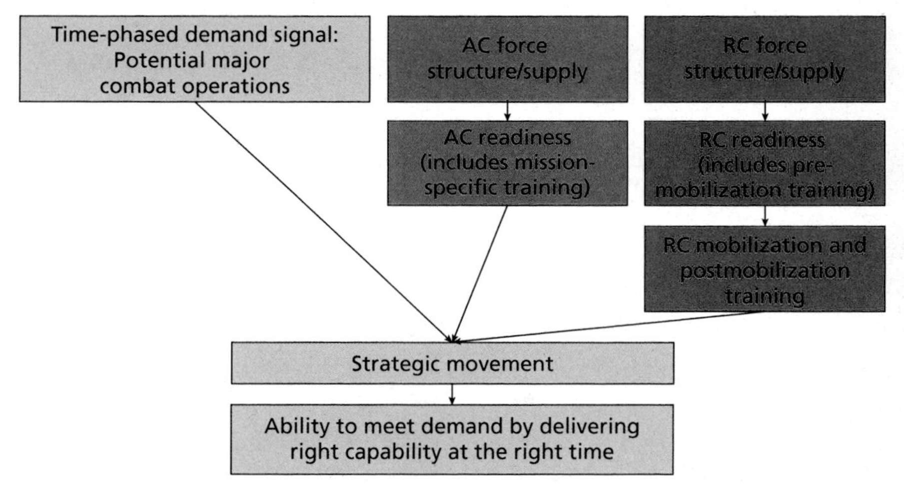

SOURCE: RAND visualization of force generation process.
RAND *RR1516-2.1*

most ready RC units will go through a mobilization process and have their readiness validated before they can deploy; most, if not all, will also conduct at least some postmobilization training and preparation, given the limited time available for training (generally, 38 to 39 days per year) when RC units are not mobilized.[1] Once AC or RC units are ready to deploy, they must go through a strategic movement process—the coordinated road, rail, air, and sea movements that deliver them into their area of operations (AOR).[2] After strategic movement to the AOR (and the reception, staging, onward movement, and integration

[1] This process culminates with validation that the unit is ready to deploy. The Secretary of the Army is required by statute to validate that Army National Guard units are ready for deployment, per U.S. Code, Title 32, Section 105. Title 10 does not specifically require the Secretary of the Army to validate Army Reserve unit readiness. However, Title 10, Section 3013, describes the Secretary of the Army's general responsibility to train and mobilize units, which can be construed as a requirement to validate Army Reserve readiness prior to deployment.

[2] In the interests of simplicity, we ignore any deployed units (AC or RC) that may be immediately available in the area of conflict and able to meet one of the TPFDD demands.

that follows it but is component neutral), the unit is "delivered" to the Joint Force Commander (JFC) to meet one of the requirements established in the TPFDD.

For the AC, one consideration for overall unit readiness is general and mission-specific training, including training required to bring units from lower levels of overall readiness to C-1 status before beginning strategic movement.[3] Similarly, RC readiness includes training activities that occur before the mobilization process starts. While our models do not address this training separately, the rate at which we assume both AC and RC units become available for deployment takes these factors into account.

Developing The Model Structure

To understand where to focus modeling efforts, we first had to identify the relevant bottlenecks in the system described in Figure 2.1. Our assessment treated the demand signal, including the timeline by which units need to arrive in theater, as exogenous. It was given to us by the scenario or CONPLAN that we used to assess the force generation process. It was important to choose a demand signal challenging enough with regard to speed and size to tax the mobilization and transportation systems we wished to investigate. A demand signal that was slow and/or small would tax neither and give us little or no analytic insight. Within those constraints, any scenario or CONPLAN demand signal (TPFDD) was acceptable for us to test and measure our mobilization models and output.

Initially, we did not assume that strategic movement was also exogenous. In fact, one question we wanted to answer was the extent to which the RC mobilization (and AC deployment) system could produce ready units that would end up "waiting" for transportation. Presumably, if these systems could provide units to the mobility system faster than the mobility system could accommodate them, then less-

---

[3] Note that some units may be deployed at less-than-C-1 status for certain missions. Still, training may be required to achieve the state of readiness required for the assigned mission.

ready RC units requiring a longer period of postmobilization training could still meet war demands; in this event, a case could even be made for migrating force structure to the RC.[4] Therefore, we began by examining mobility studies and actual TPFDD to understand how and where shortfalls in transportation could be identified. However, our research revealed that the existing TPFDD and mobility studies already represent an acceptable compromise between the JFC and U.S. Transportation Command (TRANSCOM). This compromise reflects a risk-acceptable (to the JFC) and transportation-feasible (from a TRANSCOM perspective) flow of an operationally sufficient force. Further research we conducted on strategic lift issues confirmed the availability of sufficient lift capacity with the use of various policy, contractual, and legal options for rapid access to merchant shipping. Regardless of whether the U.S. government chooses to use these various options to expand sealift capacity, the tools are in place to obtain more than enough capacity. There may be sealift capacity shortfalls in the initial weeks, but this is best addressed by prepositioning and forward deployment.[5] In any case, the RC is not well positioned to meet demands in the initial weeks. RAND's examination of the strategic lift issue appears in Appendix A.

---

[4] This is always subject to the typical "ceteris paribus" caveat. Nothing in this analysis suggests that mobilization, readiness, warning, or any of the other issues we discuss and examine should be the sole determiner of AC/RC force mix or design. There are, as the introductory paragraphs indicated, a wide variety of factors that must all be considered in making those decisions. Rather, this analysis suggests ways to factor additional considerations into the discussion.

[5] As our eventual findings suggest that mobilization output will be relatively slow at current capacity levels and current assessments about warning, this early and temporary shortfall in shipping primarily affects AC units and not RC units that have exited the postmobilization training pipeline. In the event that a range of unlikely outcomes do occur—very early preconflict mobilization, higher capacity in MFGIs, faster training of RC units at the MFGI, and intelligent sequencing of RC units through the Mobilization Force Generation Installation (MFGI)—then it is possible that RC units could end up waiting for shipping. However, if the force flow also began early (or at least mobilization and contracting of shipping assets did) then the likelihood of shipping as a constraint is again unlikely (against the agreed upon TPFDD, even if not against a more-aggressive force flow desired by the COCOM, as discussed in the next paragraphs).

It is possible (and other studies may suggest) that the feasibility of the force flow and the risk inherent in it are time specific. In other words, as strategic mobility asset inventories, unit designs, operational plans and concepts, or military technologies change, the supportability and level of risk inherent in an "old" TPFDD or mobility study may also change. However, that analysis is beyond the scope of the study. Similarly, a more-aggressive operational plan might be supportable, given a more-robust transportation inventory; however, we were not in a position to assess what might be possible in that realm.[6] Based on these factors, and buttressed by SDDC-TEA studies and other analysis,[7] we chose to also treat strategic movement as exogenous. Essentially, we inverted the initial assumption and found that (under current investment and policy options) transportation would be available as units required it for deployment—both AC units deploying from their home stations or RC units deploying after mobilization activities.[8]

Given the assumptions that both the demand signal and the strategic movement constraint can be treated as exogenous, our modeling efforts focus only on the mobilization portion of the readiness pipeline. To understand the extent of the possible contribution of the RCs to a deployment to a major contingency, we begin by maximizing RC contributions to the TPFDD. In other words, our default, for modeling purposes, was that if the mobilization system could produce an RC unit

---

6 And it is unlikely that the transportation inventory will be materially increased in the short to medium term.

7 Additional justification is based on RAND analysis of SDDC-TEA, Analysis of Mobility Platform (AMP) output data for a large-scale, time-constrained illustrative deployment. We find that the ability to deliver a large force within the time allotted by modern operational plans is highly dependent on decisions about commercial shipping usage, such as activation of the Voluntary Intermodal Sealift Agreement (VISA) program and use of contract-based commercial shipping from a variety of carriers. Given a high degree of flexibility and expenditure on shipping capacity, even very ambitious plans can be supported.

8 It would be an interesting follow-on study to determine how fast the mobilization system would have to operate in order to reverse the challenge and make transportation the limiting factor. We did not do this analysis, but the material in Appendix A suggests that if the U.S. government exercises all the options it has to expand sealift and airlift capacity, the mobilization system is unlikely to outstrip lift capacity. So the key question is which of those options are likely to be exercised in different scenarios.

in time to meet a TPFDD demand, then the sourcing system would select an RC unit. This allows us to provide analysis that is essentially a boundary—how much of the TPFDD *can* the RC support—rather than a normative assessment of how much it *should* produce. To do this, we also assumed that the AC could support any demand that the RC could not, within the limits of AC unit supply (which, in turn, was bounded by inventory and readiness assumptions) for that capability.

Understanding Postmobilization Training Times

To gather the input parameters and additional assumptions required for creating an analytic model of the conceptual framework in Figure 2.1, we conducted a literature review followed by unstructured discussions with experts from across the Army's mobilization infrastructure, including Headquarters, Department of the Army G-3/5/7 Operations Directorate (Mobilization Division); U.S. Army Forces Command (FORSCOM) G-3/5/7 Plans and Force Provider Division; Director, Army National Guard G-3 Operations and Training Directorate, Mobilization and Readiness Division; First Army G-3/5/7 (several divisions represented); Camp Shelby Director of Plans, Training, Mobilization and Security; Fort Hood Director of Plans, Training, Mobilization and Security; key leaders from the 29th Division Headquarters, Virginia Army National Guard; and key leaders from Training Division West, Fort Hood Texas. In each discussion, we attempted both to validate our findings from the literature search and to identify resource and process challenges to rapid mobilization, training, and deployment of RC forces.

The critical information we gathered from these discussions included an appreciation for speed at which the Army could expand MFGIs capacity that will set the initial conditions in the event of a large-scale mobilization, as well as an understanding of the critical factors and assumptions that affect the elasticity of that capacity.[9] We also

[9] MFGIs are the Army installations designated to provide premobilization training and support, combat preparation, postmobilization training, and sustainment capabilities to

gained an appreciation for current expectations for RC postmobilization training timelines. This is a critical variable for our analysis, but it is a difficult one to quantify with any precision in the context of an unplanned major warfighting contingency.

Ample data on premobilization and postmobilization training and preparation times for RC units that deployed to operations in Iraq, Afghanistan, and elsewhere over the past decade are available. However, these deployments were generally for known missions planned well in advance; with mission-specific training requirements focused

---

both AC and RC units.

> MFGIs will be employed as needed to generate trained and ready RC forces and to project those forces to support Combatant Commander requirements. Installations designated as Primary MFGIs (PMFGI) have the capacity to conduct mobilization, deployment, and training operations for RC and AC units, and support Joint units as required. Secondary MFGIs (SMFGI) are Active Army installations that (in addition to AC unit support) also support RC mobilization and RC premobilization training. SMFGIs have the capability to mobilize RC units during periods of surge or exceptional levels of effort. Contingency MFGI (CMFGI) are Army installations that may be used as necessary to support postmobilization training and deployment of RC units during exceptional levels of effort (Michael S. Tucker, statement to the National Commission on the Future of the Army, August 18, 2015, p. 6).

For a good accounting of the current planned capacity and throughput, see National Commission on the Future of the Army, "Force Generation Subcommittee: Mobilization Force Generation Installation (MFGI)," information paper presented at meeting, November 5, 2015. This information paper notes both that

> In late 2014, Army Enterprise Partners, Medical Command (MEDCOM), Installation Management Command (IMCOM), Army Materiel Command/Army Sustainment Command (AMC/ASC) and FORSCOM looked at current, future, and contingency mobilization requirements to determine if reduction from 3 MFGIs (Joint Base Maguire Dix Lakenhurst – JBMDL, Ft. Bliss, TX, Ft. Hood, TX) to 2 MFGIs (Bliss, Hood) was possible without incurring significant risk.

and that

> In a 25 August 15 information paper, FORSCOM indicated that a simultaneous or near simultaneous mobilization requirement of approximately 8,000 pax would require reactivation of additional MFGIs above/beyond the two (2) that are currently active. "A deliberate expansion in mobilization capacity, using current SECDEF mobilization policies and contracting procedures would require between 180 days and 225 days to complete. If statutory requirements and SECDEF policies can be waived, the time to re-activate the primary MFGI can be minimal—less than 30 days while building to maximum throughput capacity over a period of time.

on counterinsurgency and stabilization operations; in circumstances in which a substantial amount of training and preparation could be completed prior to mobilization; and in theaters where equipment was already provided, which eliminates the need for post-training maintenance of and preparation to move unit equipment. The current mobilization process is built around training units that have had months (or years) of notice of their deployed mission and AOR, so units have had time to cross-level people and equipment and arrive at the MFGI "ready to train" and already far along on both their mission-essential task list and theater-specific training requirements.

These circumstances are completely different from an unplanned crisis involving a major combat operation in Korea or Europe, which would potentially require the very rapid mobilization and deployment of a high number of units from the RC with little or no notice, possibly to execute more complex or challenging missions than experienced during the counterinsurgency and stabilization environment of recent deployments. Unfortunately, none of the officials we met with could estimate how the changed environment would affect the time required. This analysis deals with this fundamental uncertainty by using recent historical data as a (likely optimistic) baseline for the required postmobilization training time, but it also explores a range of possible postmobilization times to assess the sensitivity of the analysis to this factor.

Low-Fidelity Modeling to Identify Key Factors

Using the basic conceptualization presented at the start of this chapter and a set of baseline postmobilization training times and MFGI capacities derived from the discussions described above, we first developed a simple, low-fidelity model to gain insight into the mobilization process. This model is described in Chapter Three. The low-fidelity model combined with the research described above helped us develop three key insights worth further examination. First, early mobilization decisions or reductions in postmobilization training time improve the ability to meet demand. Second, brigade combat teams (BCTs) and combat aviation brigades (CABs) fill most (if not all) of the capacity of the RC training pipeline for extended periods. Third, there is a trade-off between the number of demands met and the size of the units trained. Specifically, because of the limited postmobilization training capacity, the system can produce a high number of small units or a low number of large units. So measurement of success may depend on whether the metric is the number of RC soldiers or the number of RC units successfully mobilized and deployed to meet wartime demands. Analysis of the operational effects of delayed arrival of specific unit types is beyond the scope of this report.

From the Low-Fidelity Model to the High-Fidelity Model

Given these initial insights, we developed a second, higher-fidelity model that would allow us to leverage the knowledge gained in the first model and more closely explore alternatives for maximizing output of mobilized RC soldiers or units. The high-fidelity model was designed to explore unit sequencing and trade-offs between training different types of units that are highlighted by the results of the low-fidelity model. This high-fidelity model provided a more in-depth look at the following issues previously examined by the low-fidelity model:

- Given a best-case scenario, how much of the demand could be filled by RC units?
- How sensitive is the mobilization system's output to the timing of the decision to begin mobilization?
- How sensitive is the output of the mobilization system to sequencing or prioritization of units through the system?
- How sensitive is the output of the mobilization system to RC postmobilization training and preparation time, which can vary by the type of unit and complexity of its mission?
- How robust are the results of the analysis? Are they overly sensitive to initial assumptions, including assumptions about AC readiness?

Findings from the high-fidelity model confirm the importance of the timing of the mobilization decision, the length of postmobilization training times, and the sequencing of units of varying size and complexity. The high-fidelity model allows us to examine the RCs' best-case contribution to demand in a large, time-sensitive contingency, given each set of input parameters and assumptions about desired unit sequencing; this lets us better understand the magnitude of the effect of these decisions beyond results from the low-fidelity model. Both the low- and high-fidelity models and the data used to drive them are detailed in Chapter Three. Chapter Four will examine the implications of this analysis and provide a set of recommendations about how to incorporate those findings into future AC/RC structure discussions.

CHAPTER THREE

Modeling Throughput

In this chapter, we describe the input data necessary for modeling postmobilization training capacity, as well as the two different throughput models (low-fidelity and high-fidelity) that were developed as part of this analysis. Both models aim to develop a better understanding of the factors driving mobilization speed for the RC in the event of a large-scale future conflict overseas. Such a conflict would require mobilizing many units with little or no notice, getting them ready to operate in a major combat environment, and deploying them in time to meet a commander's needs on the ground.

The low-fidelity model is intended to identify the relative importance of three factors driving postmobilization training throughput—existing and ready training base capacity, the timing of the mobilization decision, and postmobilization training time (with variation based on unit type and mission complexity)—and to confirm (or refute) what we learned from discussions with Army commands. The model output showed us that, in addition to these three levers, the way units of different types are sequenced in the mobilization pipeline can significantly affect results. Additionally, the low-fidelity results highlighted the importance of including assumptions on AC readiness in the modeling work. Therefore, we chose to develop a high-fidelity model to explore unit sequencing and prioritization, as well as to model for AC readiness.

The high-fidelity model generates the "best possible"[1] mobilization schedule for a particular set of assumptions. The model is designed to address the extent to which the RC could be used to fulfill demand in a best-case scenario. We find that successful employment of the RC to meet TPFDD demands depends first and foremost on the timing of the mobilization decision and secondly on unit postmobilization training time. However, best possible mobilizations prioritized smaller units with shorter training requirements that can be trained quickly and scheduled with flexibility to meet a demand schedule. Relatively few BCTs and CABs are mobilized, largely because training time for these units may be longer than the initial period of TPFDD demands.

In the rest of this chapter, we begin by presenting input data on force structure, postmobilization training times, and MFGI capacity that are common to both models. We then present low-fidelity and high-fidelity model overviews and modeling results. The analysis presented here is designed to provide representative results that can be used to explore how changes in policies and processes affect the ability to field ready RC forces; it is not intended or designed to provide a specific, prescriptive timetable for the scheduling of future training. This comparative analysis informs the discussion of future policy decisions presented in Chapter Four.

Input Data

As explained in Chapter Two, simplifying assumptions (such as the exogeneity of a transportation-feasible demand signal and the availability of sufficient strategic lift) allows us to focus the modeling work on four key parameters that affect postmobilization throughput: postmobilization training time, capacity at postmobilization training facilities, timing of the initial mobilization decision, and prioritizing which units will go through postmobilization training first.

---

[1] Loosely speaking, *best possible* can mean either "most units" or "greatest number of soldiers"—the overall conclusions are the same in both cases.

Postmobilization Training Time

Increases or decreases in postmobilization training time can result from changes in premobilization readiness, efficiencies or inefficiencies in postmobilization training, or changes in acceptable standards for ready units (e.g., must they be C-1, or is C-2 acceptable; can current COCOM-specific theater-entry requirements be waived in order to provide units more rapidly?). While these are very different options, their effect on the ability to mobilize forces is the same: Postmobilization training time will either be reduced or increased. In this way, postmobilization training time is used to represent a variety of readiness issues. Transportation times to the mobilization station, and from the mobilization station to postmobilization training locations, and from postmobilization training locations to embarkation and load locations are not modeled explicitly because they are a relatively "fixed cost" in time and represent a relatively small portion of the time needed to mobilize and prepare RC units for deployment. A planning factor to account for this sort of overhead and travel time is included in the training times used throughout this analysis; if training days required for a particular unit type were zero, we still included a training delay to account for travel.

Capacity at Postmobilization Training Facilities (and How Quickly It Can Be Expanded)

Our discussions with Army commands made clear that training capacity could be represented by the available bed spaces at the various training sites. While the availability of trainers in general, trainers of the right type, and multipurpose training brigades that can accommodate a variety of units are also potential bottlenecks in the postmobilization training process, discussions with Army organizations revealed many options and work-arounds for obtaining trainers and indicated that bed space could be used as a proxy for the set of facilities and services that are the limiting factor in the current environment. Bed space, and the set of support services for which this is a proxy, currently limit both the initial capacity at MFGI facilities and the ability to expand capacity at in-use facilities or open unused facilities. Therefore, bed space is, in this context, synonymous with installation capacity. A related constraint is that postmobilization training of BCTs and CABs can only occur at certain sites and is limited to one BCT or CAB at a time per site. The current set of available MFGIs and assumptions about their opening speed are discussed later in this chapter.

Timing of the Initial Mobilization Decision

Because ramping up capacity at postmobilization training facilities takes time and training times are long for some unit types, beginning the mobilization process earlier can have a significant effect on the ability to create ready units in time to meet a deployment timeline. In the event that sufficient early warning is available and the political process allows it, the option to begin mobilization well ahead of the actual force flow timeline could provide additional flexibility to policymakers.

Prioritizing Which Units Will Go Through Postmobilization Training First

The sequencing of units of different sizes and with varying mission complexity (and therefore varying postmobilization training times) has an effect on the throughput of the overall postmobilization training system. This idea came up in our discussions with Army organizations, but its importance only became clear after the initial modeling work. As discussed later in this chapter, the trade-off between first training smaller, quicker-to-train units versus larger units with very complex missions completely changes the way in which MFGI capacity is allocated.

Key Data Inputs for Modeling

To study these factors, both models require data on RC unit inventory (including unit size based on the number of soldiers authorized), postmobilization training time, and a demand signal specified by unit type and required timing of arrival in theater. We used inventory and authorization numbers for the AC and RC obtained from the Army Equipping Enterprise System in December 2015, a demand signal derived from TPFDD, and baseline training times derived from First

Army training plans.[2] Although it has been aggregated and simplified for modeling purposes, our demand signal is modeled on actual TPFDD for a large Army deployment—but does not directly represent it in terms of forces or force flow data. In an effort to present analysis at an unclassified level, only results aggregated across time and unit type are presented in this report.[3] Conclusions drawn from this analysis are consistent with and as informative as the more-granular classified results. Training times include several days above planning factors to account for administrative issues and travel.

To simplify modeling and ensure that outputs are readable, we aggregate RC unit types into unit categories based on similar numbers of authorized soldiers and postmobilization training times. Our models allocate MFGI capacity to these unit categories to meet demands, which are also aggregated up to the same level. We assume that planners will make efficient standard requirements code (SRC)-level MFGI assignments in a real-world mobilization.[4] BCTs are tracked based on their maneuver battalions, using an AC 490,000–based force with two-battalion BCTs (with the exception of Stryker BCTs, which have three battalions) as a model.[5] CABs are tracked based on air cavalry squadrons (OH-58), attack/reconnaissance battalions (AH-64), and attack/reconnaissance squadrons (OH-58D). A single arrival date estimate for these large units is derived from the arrival dates of each unit's largest

---

[2] As described in Chapter Two, these training plans reflect postmobilization training times in a counterinsurgency/stability operation environment, in which a substantial amount of preparation has been completed prior to mobilization.

[3] Additional detail at the classified level is available upon request.

[4] The Army uses SRCs to distinguish units by type; SRCs are the basis for the Modified Table of Organization and Equipment (MTOE) designation of a unit.

[5] Our models track BCTs in terms of the approximately 600-person maneuver battalions. While only these components of the BCT are explicitly modeled, model logic forces these units to go through postmobilization training together as a whole BCT. Additionally, model logic allocates sufficient MFGI training capacity for all of the smaller components of the BCT to train alongside the maneuver battalions at the same MFGI at the same time. Furthermore, our modeling is based on a two-battalion BCT, and therefore likely understates the demand capacity required by the BCT in cases where a separate Army National Guard maneuver battalion is mobilized with an Army National Guard BCT to create a full three-battalion BCT.

components. The illustrative demand signal for the aggregate unit types includes a force flow spanning 20 weeks and approximately 1,300 units (at the aggregated unit type level) consisting of 176,000 soldiers. Illustrative demands are assumed to reflect "ready-to-load" dates (the date that a unit is ready to load onto its strategic lift platforms; in this case, primarily onto ships).[6] In other words, we used the ready-to-load date for personnel as the time by which a unit must have completed its postmobilization training and administrative activities in order to meet the strategic transportation demands that ensure the unit makes its TPFDD-directed LAD.

In addition to supply data, our models rely on a simplified representation of the availability of, and capacity at, First Army's postmobilization training facilities.[7] Based on discussions with First Army and FORSCOM, as well as First Army mobilization planning documents, we make several assumptions. We assume that training takes place across eight major installations, two of which (Fort Bliss in Texas and Camp Shelby in Mississippi) are available for training BCT components, and one of which (Fort Hood in Texas) has capacity for training CAB components. While the RCs may fall in on additional training facilities vacated by AC units, we do not account for this extra capacity in our analysis; because of the demanding nature of the type of short-warning, major contingency operation of interest to this study, it is highly unlikely that AC training capacity would become available to the RC. The eight training installations are not all fully available on the first day of the next conflict; we assume that Fort Hood and Fort Bliss provide some training beds without warning, while Joint

---

[6] For more information on the dates generally used in a TPFDD, see James C. Bates, "JOPES and Joint Force Deployments," PB700-04-3, *Army Logistician*, Vol. 36, No. 3, May–June 2004, pp. 30–35.

[7] First Army is headquartered at Rock Island Arsenal, Illinois. Its mission statement reads:

> First Army, as FORSCOM's designated coordinating authority for implementation of the Army Total Force Policy, partners with [Army Reserve] and [Army National Guard] leadership to advise, assist, and train RC formations to achieve Department of the Army directed readiness requirements during both pre mobilization and postmobilization through multicomponent integrated collective training, enabling FORSCOM to provide Combatant Commanders trained and ready forces in support of worldwide requirements (see Headquarters, First Army, "Mission," undated).

Base McGuire-Dixon-Lakehurst (JBMDL) in New Jersey, Camp Atterbury in Indiana, Joint Base Lewis-McChord (JBLM) in Washington state, Fort McCoy in Wisconsin, Fort Stewart in Georgia, and Camp Shelby come online after some time. As shown in Figure 3.1, we assume

Figure 3.1
Training Capacity

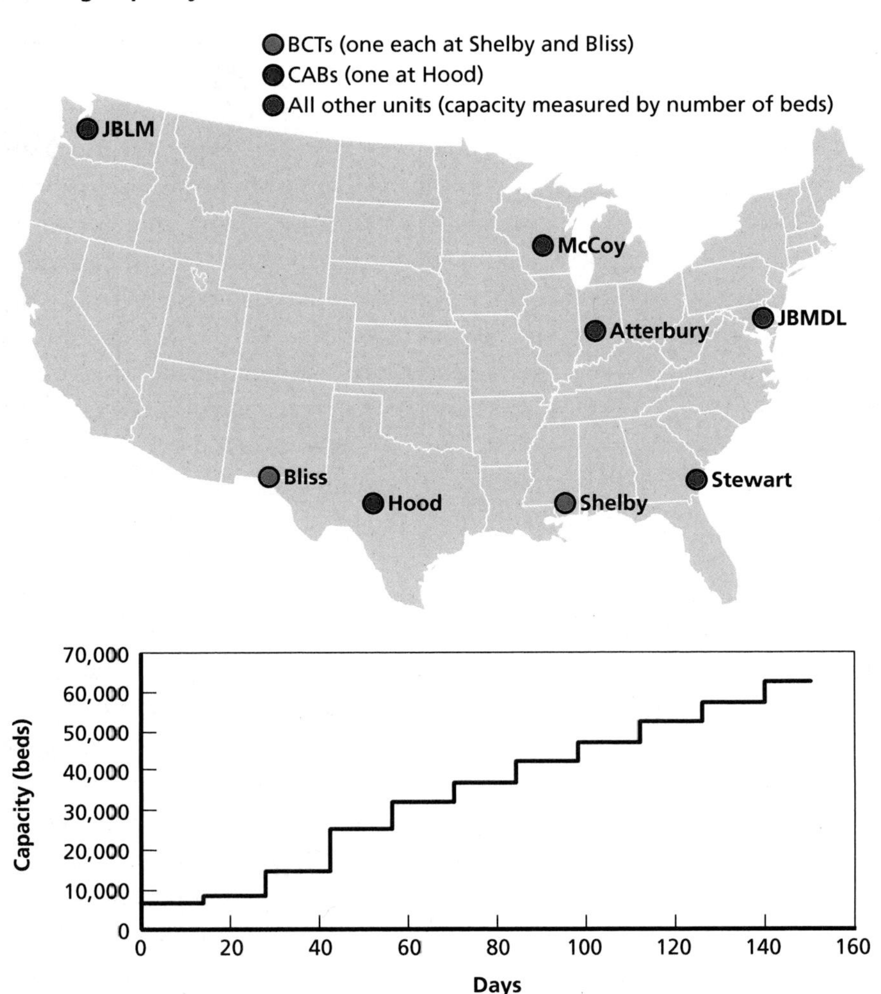

SOURCE: Based on RAND analysis of FORSCOM and First Army data and planning factors.

RAND *RR1516-3.1*

that capacity at the bases increases every two weeks, following a ramp-up schedule. Full operational capacity for postmobilization training is around 66,000 beds. Because training must occur as a unit, both the low- and high-fidelity models require enough available bed space for an entire unit before a unit is allowed to train.

Consistent with First Army mobilization planning documents and interviews with mobilization planners at First Army and the MFGIs, we assume that the number of MFGI facilities available for RC postmobilization training, as well as the total capacity at these facilities, is basically fixed. It would be very expensive to convert additional facilities for use as MFGIs, and there is no plan in place for doing so. AC facilities would be training AC units in the case of a major combat operation like the one used in this example. Instead of exploring changes to the available MFGI capacity, we vary assumptions about the rate at which capacity at this set of MFGIs becomes available at the start of a conflict. First Army sources indicate that expansion of MFGI capacity will resemble a step function (some expansion at set periods of time) rather than a continuous or linear function (some expansion every single day). We chose two weeks as an example step function based in large part on discussions with MFGI staff about the activities and tasks needed to increase capacity. These tasks include the need to establish contracts for a variety of services and to allow the contractors time to hire, train the appropriate labor, and begin delivery of services. Contractors cover a variety of nontraining support activities, ranging from providing food, fuel, maintenance, and network infrastructure to preparing buildings for occupation and providing transportation services. Mobilized soldiers or contract workers may provide additional services in the areas of administrative and medical processing of mobilizing soldiers and in the coordination and management of moving mobilized soldiers and their equipment by road, rail, and air from mobilization station to MFGI and then from MFGI to ports of embarkation (air and sea). The two-week step function represented in Figure 3.1 is an optimistic representative expression of our understanding of the First Army planning factors and the speed with which they can expand capacity.

There are additional training restrictions for BCT and CAB components because they need special ranges and training facilities. As mentioned, only Camp Shelby and Fort Bliss have the infrastructure capable of supporting BCTs, while CABs can only train at Fort Hood. The model reflects this constraint; only one RC BCT or RC CAB, including all their subcomponents, can train at each of these bases at a time, and no non-CAB and non-BCT components can train at these bases if they are used for CAB or BCT training.

Our models explore three options each for unit postmobilization training time and rate at which MFGI capacity becomes available. The varying training time and capacity assumptions are designed to explore the effect of changes to training and facility policies on postmobilization training throughput. Variation in postmobilization training time from currently planned baseline levels could occur in a number of circumstances. For example, postmobilization training time might be reduced if additional training requirements could be addressed during premobilization training activities. However, one should note that our baseline postmobilization training times are already based on historical models with a substantial amount of training occurring prior to mobilization. Shifting even more training to premobilization would likely require an increase in premobilization training days and associated funding. Additionally, this could place a significant burden on part-time RC soldiers. Postmobilization training time could also be reduced through changes in requirements, such as accepting risk in some areas or tailoring training to a specific mission. Improvements to specific postmobilization training activities could also theoretically increase efficiency and allow those with similar skills to be trained more quickly, although the Army has made substantial improvements in training efficiency over the past decade. Because our postmobilization training window includes time for travel, as well as medical and administrative activities, efficiencies in these areas could contribute to less time spent at the postmobilization training facility. On the other hand, preparing for a major combat operation may impose longer or more complex training requirements than units experienced in preparing for counterinsurgency and stabilization-focused operations over the past decade. Because of the high number of potential policies, requirements, and practices that could affect postmobilization training time, our model runs explore large variations in training time requirements that are both shorter and longer than our baseline. However, in some cases, even the lengthiest estimates may underestimate the time needed to finish required postmobilization training, because training is both mission- and threat-dependent and the specific training requirements for a large crisis may not be known ahead of time.

Our models also allow for variation in the speed at which postmobilization training capacity (number of beds) becomes available based on analysis of the availability and level of staffing and equipment at the different installations. Ramp-up curves differ depending on policy decisions regarding which facilities are kept in a warm or semi-warm status (meaning different levels of minimal maintenance intended to facilitate and expedite reopening of these facilities) during the steady state, in anticipation of a major future mobilization. Variation in which bases are kept warm, the number of beds that become available immediately upon opening a base for training, and the rate at which additional beds can be brought online after initial base opening affect the ramp-up curves. Our models assume that only Fort Hood and Fort Bliss are kept warm prior to a mobilization decision, and that these two facilities operate below peak capacity during peacetime. In other words, all of our model runs assume the same level of MFGI capacity at the start of mobilization operations. All excursions in capacity ramp-up speed explore only changes to the rate at which other bases can be brought online and ramped up to full capacity.

The baseline training time used in our models is derived from First Army and FORSCOM planning factors. We then calculate a shorter (50 percent of baseline training days) and longer (120 percent of baseline training days) training time for each unit type. As mentioned earlier, the baseline capacity ramp-up schedule is based on current First Army planning documents and follow-on conversations with First Army and FORSCOM. Because the ramp-up schedule for base capacity shown in Figure 3.1 reflects a fairly optimistic estimate given a cold start without warning, we also include excursions using two slower ramp-up speeds. Alternative capacity ramp-up curves are shown in Figure 3.2. These curves are based on subject-matter experts' estimates of the rate at which capacity could become available at each facility.

Figure 3.2
MFGI Capacity Ramp-up Options

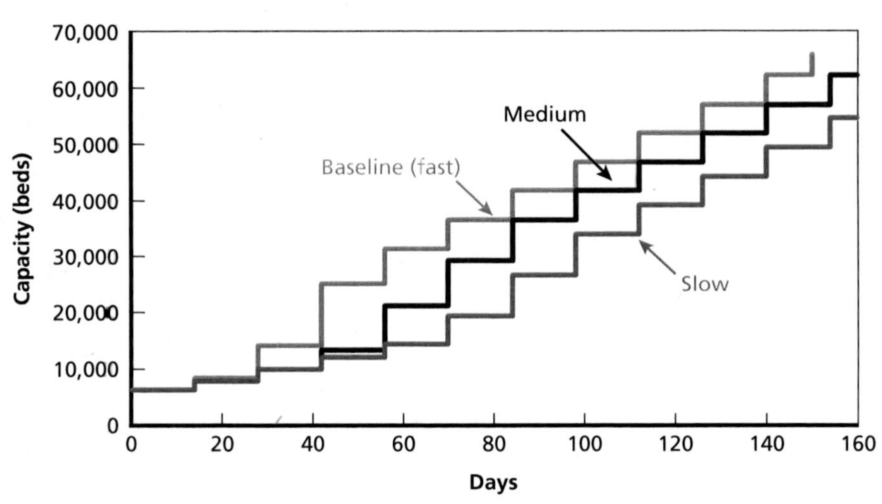

SOURCE: Assumed ramp-up curves based on RAND analysis of First Army data.
RAND *RR1516-3.2*

Low-Fidelity Model

We first developed a low-fidelity model intended to identify primary factors driving postmobilization training throughput (the number of ready RC units and soldiers produced by the system) and confirm (or refute) what we had learned from discussions with Army commands. The low-fidelity model explores alternate future environments defined by variations in MFGI capacity ramp-up speed and unit-level postmobilization training times. The model examines the relative effect on RC postmobilization training facility throughput of these broad policy levers without making any intelligent scheduling decisions. This simple model revealed the need for more granular analysis that explores prioritizing different types of RC units and highlighted the importance of including assumptions on AC readiness in the modeling work.

We use the Python programming language to implement a discrete event simulation model of unit mobilization and training.[8]

---

[8] Python Software Foundation, "The Python Language Reference," version 3.4, undated, last updated June 25, 2016. Our simulation makes use of SimPy, a discrete-event simulation

Required model inputs consist of a list of RC units, the time needed to mobilize and train to the required level of readiness, the day the unit is required to have completed training and be available to deploy, and the size of the unit (number of soldiers). The low-fidelity model does not capture differences between unit types. Instead, it aggregates the unit types described earlier into three categories based on their training base requirement, which is CAB, BCT, or "other."

The selection of units for training is based on a "first-out, first-in" algorithm, meaning that those units with the earliest demand signal are sent into the training pipeline first. The model sorts units to determine which are first in line. There is an optional user-specified parameter for a unit's importance. If this parameter is specified, the model sorts the units based on the user's specified order of importance. Otherwise, the model sorts the units by the day on which they are required. We use this option to begin exploring prioritization between different unit types.

The model assigns each unit, in order, to a specific pipeline based on its unit type, and then the model checks to see if a base has enough bed spaces (a proxy for installation training capacity) available. If open bed space is available, the unit occupies that base for the length of its combined mobilization and training time. The low-fidelity model does not force all personnel from a particular unit to train at the same training base; because bases are large, we assume planners could easily remedy such issues in the real world. As outlined in the next section, the high-fidelity model alleviates this shortcoming. In the event not enough bed space is available, either because a unit is already occupying the beds or because additional beds have not been created, the model places the units in a queue for the next available beds. When CABs and BCTs from the RC are training, the model does not allow space at Camp Shelby, Fort Bliss, or Fort Hood to be used for units other than CAB or BCT components. This implies that capacity at these facilities is allocated to CABs and BCTs, even if these units cannot be trained

---

library built for Python by Ontje Lünsdorf and Stefan Scherfke. SimPy has an MIT License is available online (see Python Software Foundation, "SimPy 3.0.10," undated, last updated September 1, 2016).

in time to fulfill a demand. Because Fort Bliss and Fort Hood are the only two warm MFGIs at the start of all model runs, all of their available capacity is consumed by one CAB and one BCT if they are mobilized first. Real-world planners would prevent scheduling decisions that allocate space only to units that cannot be mobilized in time, while the model does not have such checks. For these reasons, the low-fidelity model may underestimate the ability to meet demand.

A unit is declared available once it has passed through the pipeline and completed its required training time. From there, the model checks whether that unit can fill any demands; if there are any unfilled demands for this type of unit, the model deploys the unit. If no RC units are available to fill a demand, the model will slot an AC unit, with no restrictions on AC inventory or readiness. The effect of AC inventory and readiness are explored in more detail in the high-fidelity model. Figure 3.3 provides a graphical representation of this process.

As an example, consider an armored BCT (ABCT). Because of its unit type, an ABCT must go to Camp Shelby or Fort Bliss. If these bases are not yet online, the ABCT must wait for one of them to open with sufficient capacity. Once the base is open and has enough beds available to accommodate an ABCT with all its subcomponents, the ABCT enters and begins postmobilization training. After training, the ABCT becomes available, and the model checks the demand signal to see if there are any demands that the ABCT can meet. The model slots the ABCT into the first demand it finds. MFGI training capacity becomes available for the next unit as soon as the ABCT finishes training, regardless of whether it is immediately deployed.

Because of its focus on macro-level trends, the low-fidelity model has some key shortcomings. Notably, it is deliberately naïve; the model generally assigns training space on a first-demanded, first-trained basis and does not optimize for efficient allocation of training space. Large units can occupy bed spaces and prevent other units from training. In addition, data on individual unit types are aggregated. Smaller units with low training times may be unfairly sidelined because they have been rolled up into a larger unit category with a longer training time. Finally, this model does not allow any exploration of the effect of AC

Figure 3.3
Logic of Low-Fidelity Simulation Model

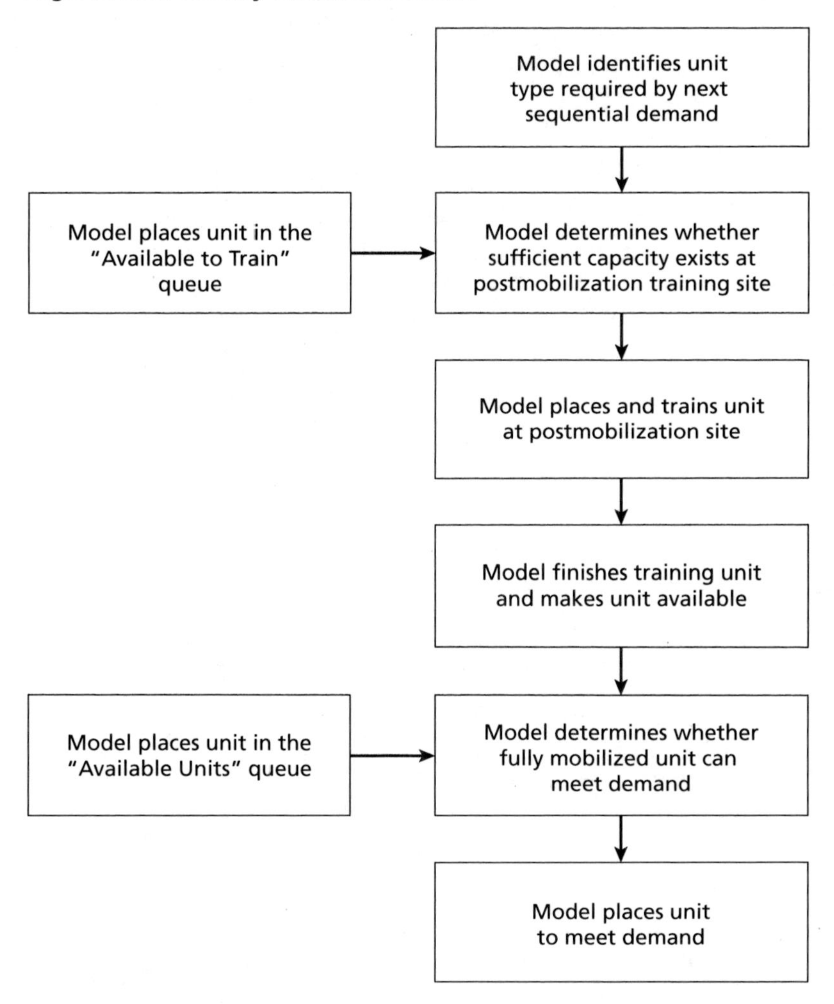

SOURCE: RAND analysis and model documentation
RAND *RR1516-3.3*

inventory and readiness. We create a second model, described in the next section, to address some of these shortcomings.

Results

We use the low-fidelity model to explore three key parameters: MFGI capacity, the timing of the mobilization decision, and changes in training time. We also use the low-fidelity model to take a first look at the importance of prioritization and sequencing, by testing the effect of excluding CABs and BCTs from the training pipeline. Figure 3.4 summarizes the results of 12 different model runs exploring these parameters against an illustrative demand signal.

We focus on the number of units supplied by the RC throughout most of our analysis, because units are the level at which COCOMs make requests for forces. While the number of soldiers can also be used as a metric of throughput, it is the fully mobilized units that are required by the force flow. Both metrics are explored throughout this report. When BCTs and CABs are excluded entirely from the model, we assume the capacity at otherwise dedicated BCT and CAB training facilities becomes available for the other units. Capacity at different facilities is assumed to be fungible across all other unit types, as indicated through our discussions with subject-matter experts.

Figure 3.4
Low-Fidelity Model Results

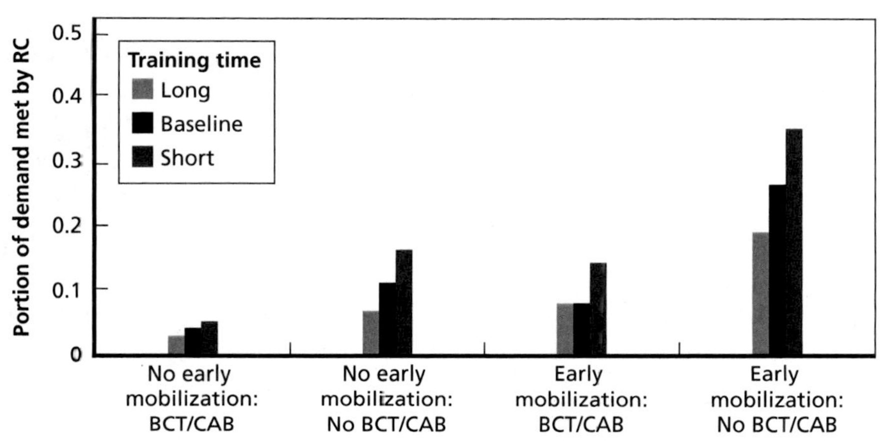

SOURCE: Outputs from RAND's low-fidelity simulation model.
NOTE: Baseline capacity ramp-up curve from Figure 3.2 is used for all runs in this figure.
RAND *RR1516-3.4*

Across model runs, we see that the RC can meet more unit-denominated demand when BCTs and CABs are excluded from the pipeline, and they can meet over twice as much demand when a mobilization decision is made eight weeks prior to the start of the conflict instead of at the start of the conflict.[9] Many real-world scenarios may not provide sufficient warning to allow for such an early mobilization decision. The eight-week window includes additional time for training capacity ramp-up and for actual training, meaning that the first units begin training as early as eight weeks prior to the start of the conflict.

In addition, the effects of an earlier mobilization decision and shorter training time are greater when BCTs and CABs are excluded from the model; when BCTs and CABs are included, training capacity at three of the MFGIs (Fort Hood, Fort Bliss, and Camp Shelby) is consumed with BCTs and CABs throughout the eight weeks of early mobilization and the rest of the training window. Even under the longest mobilization timeline, relatively few RC BCTs or CABs can be mobilized and made ready in time to meet TPFDD requirements. While only results using the baseline ramp-up curve (see Figure 3.2) are illustrated in Figure 3.4 in order to keep the chart readable, the findings above hold across all of the capacity ramp-up options explored.

Because of the size and relatively long postmobilization training requirements of BCTs and CABs, giving priority in mobilization and postmobilization training to smaller units that are quicker-to-train instead of BCTs and CABs allows the RC to meet more demands in terms of number of units. In our illustrative demand signal, the latest demand for a CAB is prior to the 16-week point. Because 16 weeks of training are required, it is impossible to train a CAB in time to meet demand without an early mobilization decision. The last demand for a BCT occurs after the 12-week mark, so it is possible that the RC could supply the last BCT demand with a 12-week training requirement.

---

[9] A period of eight weeks is selected as the shift in the timing of a deployment decision because it is a relatively short amount of time, but sufficiently long to show significant improvement. We explored alternatives, both shorter and longer. Shorter shifts provide smaller benefits, while longer shifts further increase ability to meet demand with reserve forces.

Pushing large units through the postmobilization training system can "clog" the entire pipeline.

To summarize, we can conclude that training time, which is tied to unit size and complexity, is a key driver of mobilization throughput in terms of the number of units mobilized. Given current training time and MFGI capacity assumptions, the RC can either mobilize a few larger units that can contribute closer to the end of the initial TPFDD, or a higher number of small, quicker-to-train units earlier in the fight. Using some facilities exclusively for CAB and BCT training delays other units from entering the training pipeline. Depending on the goals for employing RC units, reallocating existing MFGI space to different unit types, adding additional warm MFGI space or increasing ramp-up speed for capacity, or investing further in premobilization readiness could increase the ability of RC units to supply needed capabilities early in a fight. Increased investments in CAB and BCT readiness could help reduce postmobilization training time and free up facilities for use by other units. Similarly, allocating training resources consistent with the goals of employing the RC can have a significant effect on the number and types of demands the RC is able to meet. For example, maintaining select early-deploying RC units at higher levels of readiness shortens their postmobilization training time requirements. Examining the feasibility of such changes, including considerations of cost, policy, and political feasibility, is beyond the scope of this study. Part-time RC soldiers may not be able to accommodate an increase in training days, and sustaining readiness may be an issue when personnel turnover is high, for example. The second model further explores the trade-offs associated with different sequencing and prioritization of unit types.

High-Fidelity Model

The second model developed for this effort further explores the issues of prioritization and trade-offs between training different types of units that are highlighted by results of the first model. This is an optimization model that generates the optimal postmobilization training schedule with a particular set of assumptions. This model is designed to address the question of how much the RC would fulfill demand in a best-case scenario. Additionally, this model makes assumptions about AC readiness but allows us to test the robustness of our findings to changes in these assumptions.

The model reveals that the ability for the RC to meet a large amount of demand, as measured by both number of soldiers and number of units, depends largely on the timing of the mobilization decision, with training time assumptions also playing a significant role. Differing the ramp-up speeds for additional MFGI capacity has a lesser effect. Relatively few large RC maneuver units (BCTs and CABs) can be trained in time to contribute to the 20-week demand signal, especially without an early mobilization decision. This means that RC contributions are focused primarily on smaller units with shorter training times. Additionally, this model includes assumptions about AC readiness and tests the robustness of our findings to changes in these assumptions.

Broadly speaking, the objective of the model is to minimize the amount of unmet demand in the TPFDD. Demand can be met by an RC unit, given sufficient time and capacity to complete postmobilization training for this unit. If an RC unit cannot meet a demand, this "RC-missed" demand may be met by an AC unit. The model de facto optimizes the sequencing of RC units through the training pipeline subject to constraints related to postmobilization training time, capacity, and available RC and AC units. The mechanism for making tradeoffs between AC and RC units, and limiting the inventories of each, is described further in the next section. A formal representation of the optimization model is provided in Appendix B. Next, we describe the objective function and constraints qualitatively.

Optimization Goals

The goal of each run in the high-fidelity model is to minimize demands missed by the RC. Mathematically, the objective function minimizes the sum of the product of the RC-missed demands and their respective penalty coefficients. An *RC-missed demand* is defined as a demand that cannot be met by a trained RC unit at that time. However, the penalty is reduced if that RC-missed demand can be then met by an AC unit. This logic ensures that the optimization model "prefers" to miss RC demand for a unit type that has plenty of AC inventory available (all other things being equal), thus reducing demand that is completely unmet. Due to this mechanism, a significant change in AC/RC force mix would have an effect on modeling results. AC units do not take up capacity at postmobilization training bases. To account for AC readiness and that every AC force cannot all be ready for deployment, we assume that no more than 50 percent of the AC inventory for each unit type is ready to meet demand on the required date in our illustrative demand signal. This may overstate the AC readiness for many unit types, and we do not suggest that such a level of AC readiness is necessarily likely in a future conflict. Changing this assumption has significant effects on the total demand met but does not change our conclusions about relative participation of the RC, which is the focus of this study. We provide supporting information about this finding in Appendix C.

General Constraints

The model has a high number of constraints related to capacity and training time needed for RC units. While Appendix B provides the detailed constraint formulation for the mixed-integer optimization model (including logic constraints to ensure "if-then" and "either-or" features), here we summarize the intent of those constraint formulations, including the following:

- any demand that cannot be met by a trained RC unit (that has not met any prior demand) is classified as RC-missed
- any demand prior to an RC unit's minimum training time is considered RC-missed
- the number of RC soldiers in the training pipeline at any time cannot exceed the capacity of the training facilities at that time
- the total number of trained RC units of type *i* cannot exceed the RC inventory of that unit type
- the number of AC units of type *i* used to meet demand cannot exceed either (1) the missed demand of that type or (2) 50 percent of the total AC inventory.

We do not include additional constraints on either AC or RC inventory to account for other readiness issues. While it is clearly unrealistic to assume that 50 percent of the AC force and 100 percent of the RC force is ready for postmobilization training on day one of a future conflict, the illustrative demand signal used constrains the rate at which units will be deployed. We essentially assume that the illustrative demand signal is feasible from a readiness standpoint, at least outside of RC postmobilization training.

BCT- and CAB-Specific Constraints

In addition to these general constraints, the mixed-integer optimization model contains a series of BCT- and CAB-specific constraints based on training requirements revealed during our interviews with FORSCOM. Specifically:

- Aviation units assigned to CABs can only train at Fort Hood.
- If any CAB-assigned aviation unit is training at Fort Hood, then only other CAB-assigned aviation units may train at Fort Hood during that training interval.
- A BCT component unit can train only at Camp Shelby or Fort Bliss.
- If any BCT component unit is training at one of those two locations, then that location can only train other BCT component units.
- No more than three BCT maneuver battalions can be trained at the same time at each Camp Shelby and Fort Bliss (six total).
- No more than two maneuver CAB pacing units (see below) can be trained at the same time at Fort Hood.

One key assumption we make is that maneuver battalions—as the pacing units for a BCT—should be trained only if an entire RC BCT or CAB could be deployed to meet a demand. This means that armored and infantry BCT battalions must be trained in pairs,[10] while

[10] Note that RC BCTs consist of two battalions. The Army's reorganization of many BCTs into a three-maneuver battalion design is not yet fully reflected in scenarios and plans with detailed TPFDD data. Regardless, the change in design may shift some of the parameters

Stryker BCT battalions must be trained in trios. Similarly, we assume that RC CAB components will deploy in a configuration similar to that of an AC CAB. Therefore, at least two CAB pacing units (air cavalry squadrons [OH-58], attack/reconnaissance battalions [AH-64], and attack/reconnaissance Squadrons [OH-58D]) must be trained in order to meet a demand.[11] Again, when any CAB or BCT component is training at a CAB- or BCT-dedicated training base, no other units may train there. Because the CAB- and BCT-dedicated training bases have sufficient bed space for an entire CAB or BCT, we can assume that the CAB and BCT components not explicitly tracked in the model will have sufficient space to train along with their tracked counterparts.

To maintain a manageable number of discrete decision variables, the mixed-integer optimization makes decisions at the weekly level. All training times, MFGI capacity time series, and demands for units are rounded to the week level. The model assumes that no units are trained beyond those required to meet a future demand, even if excess capacity is available.

Results

Using the high-fidelity model, we explore the effects of three key parameters: capacity, training time, and the timing of the mobilization decision. The high-fidelity model is an optimization model, and therefore also explores the effect of optimal unit sequencing, eliminating clogs to

---

of what our model examines (number of bed spaces occupied at a BCT-generating MFGI, length of time to train a BCT, LAD for the BCT, and other factors), but it should not make our inferences invalid. In fact, if one supposes that a three-battalion BCT is larger and more complex than the two-battalion design, the inference would be that it will actually take longer to train and require more resources. This suggests an even stronger conclusion with regard to the effect of early introduction of RC BCTs into the mobilization pipeline.

[11] The concept of a "pacing item" is used in Army readiness reporting to highlight those one or two systems which are the primary basis for the combat power or mission capability of the unit; for example, the tank in a tank battalion. We have adapted this concept to allow us to simplify our analysis of BCTs and CABs. We assess that the critical combat power (the pacing item) of a BCT is its maneuver battalions, and that of a CAB is the combat battalions (attack or attack/reconnaissance) within the CAB. Therefore, we require a BCT or CAB to have these battalions present for training in order to meet a demand, as we described earlier.

the training pipeline noted in the discussion of the low-fidelity model. The analysis builds on the results of the low-fidelity model, exploring each parameter in more detail. Figure 3.5 provides a visual representation of the portion of demand that can be met by the RC under baseline assumptions, when training space is allocated optimally.

The baseline case includes medium training time based on First Army templates, a speedy ramp-up curve for training capacity (the "Baseline [Fast]" case in Figure 3.2), and a mobilization decision that coincides with the start of the conflict. The filled area represents the

Figure 3.5
Baseline Optimization Results

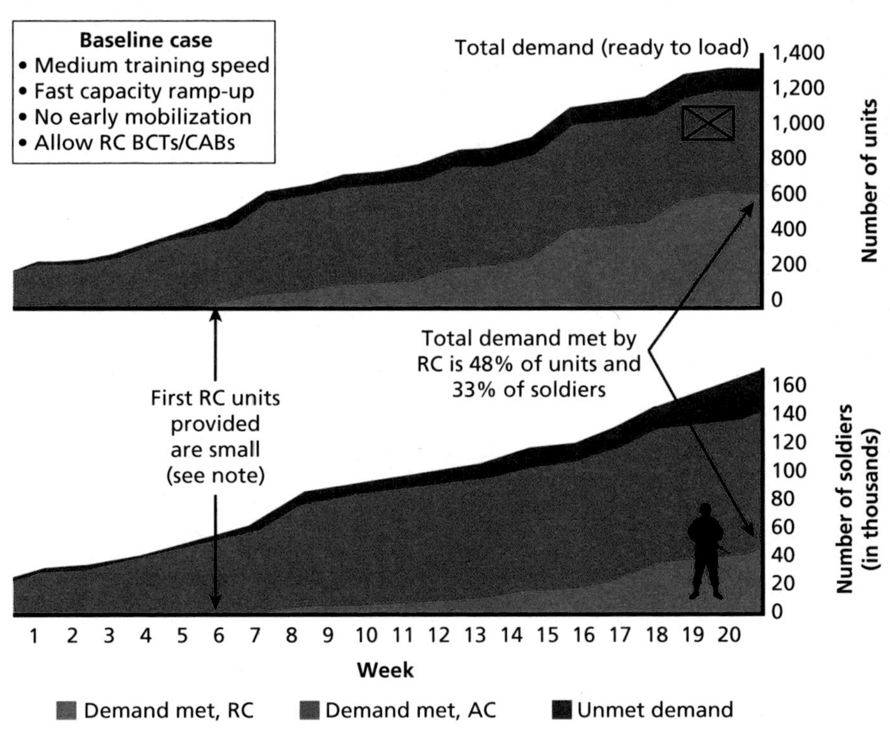

SOURCE: Outputs from RAND's high-fidelity optimization model.
NOTE: First RC units provided are small in terms of number of soldiers. Among aggregated unit types used for modeling, shortest training time is five weeks. At the more granular unit level, the RC could meet earlier demands for small units (two to three weeks). Does not account for any RC units that were in mobilization pipeline and diverted to a new contingency.
RAND *RR1516-3.5*

(cumulative over time) total demand in our illustrative signal, with yellow indicating demand met by RC units, green indicating demand met by AC units, and red indicating unmet demand. As stated earlier, a demand is considered unmet unless a unit is immediately available to meet this demand signal. Most of the red area can be filled if we allow demand dates to shift slightly to the right (one to two weeks). The top half of Figure 3.5 shows these results from maximizing RC usage on a per unit basis, while the bottom shows maximization of the total number of RC soldiers used to meet demand. Comparing these figures shows that RC units able to fill demand tend to be smaller units; the RC is able to provide 45 percent of units required in our signal but just 33 percent of soldiers. We also see that the earliest RC units are employed around the six-week mark. This is an artifact of our unit aggregation methodology, which does not yield any aggregated unit types with training times of less than five weeks. In reality, some small RC units could be trained more quickly, but their numbers in both units and soldiers are limited and do not affect the validity of the takeaways from the analysis.

Impact of Changes in MFGI Ramp-Up Capacity

From this baseline run, we explore our three parameters: MFGI capacity ramp-up speed, RC postmobilization training time, and the possibility for an early mobilization decision prior to the start of actual deployment operations.

First, we explore capacity. To show the results of multiple runs on a single plot, we use the visual framework shown in Figure 3.6.

Figure 3.6 compares results of the baseline run in Figure 3.5 with those of the two alternate facility ramp-up curves (i.e., the "Medium" and "Slow" cases) displayed in Figure 3.2. The three circles in Figure 3.6 correspond to the three separate model runs (baseline case with fast ramp-up of training capacity, baseline case with medium ramp-up, baseline case with slow ramp-up). The axes show the amount of demand met in terms of numbers of units. The horizontal axis plots the total percentage of unit demand met by the RC, while the vertical axis plots the total percentage of unit demand met by the AC and the RC together. Therefore, a result that appears higher and to the

Figure 3.6
Effect of Changes in Capacity Ramp-up

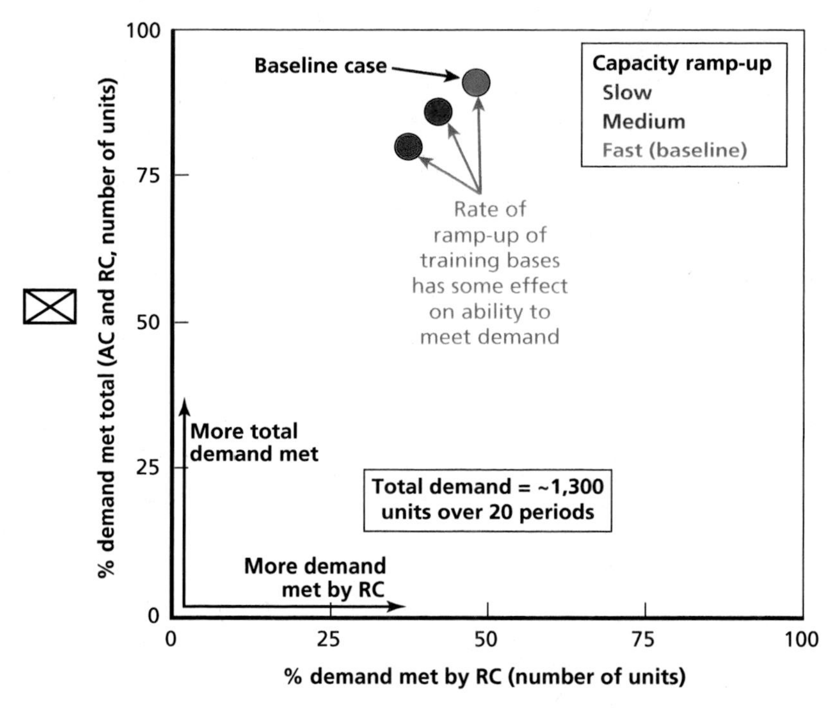

SOURCE: Outputs from RAND's high-fidelity optimization model.
RAND *RR1516-3.6*

right within the plot area is generally "better," with higher placement indicating a higher percentage of overall demand met, and placement to the right specifically signaling that the RC is able to meet a higher percentage of total demand. Any area above a plotted point represents demand that is filled by neither the AC nor the RC, indicating a shortfall under the assumptions of that run. The change in color indicates varying capacity ramp-up curves, with purple representing the baseline case. We see that slower ramp-up curves (blue and orange circles) reduce the percentage of demand met by the RC, as well as the percentage of demand met overall.

Changing assumptions regarding MFGI capacity ramp-up speed affects both the total demand met and the demand met by the RC. Across the three cases considered here, the percentage of demand met by the RC (measured in numbers of units, across the whole 20-period demand signal and across all unit types) varies from about 35 percent to almost 50 percent, while total demand met varies from about 80 percent to more than 90 percent of total demand. Across the same three cases, the percentage of demand met by the RC in terms of number of soldiers varies from about 24 percent to 33 percent, while the total demand met varies from 79 percent to 88 percent.

Effect of Changes to RC Postmobilization Training Times

The next figure expands on the results presented in Figure 3.6 to add another dimension to the set of assumptions explored. Figure 3.7 compares the same baseline run portrayed by the purple circle in Figure 3.6 with excursions using varying postmobilization training times instead of MFGI capacity ramp-up speeds. Similar to how the change in colors indicated different ramp-up speeds in Figure 3.6, different shapes are used to represent the varying options for training time explored in Figure 3.7. The purple circle again represents the baseline case, including baseline training time assumptions. The purple triangle represents the baseline case with shorter training times (50 percent as long as the baseline case training times), while the purple square represents the baseline case with longer training times (120 percent as long as the baseline training times). Different durations for training times could result from a variety of factors: varying investments in premobilization readiness (e.g., either greater or fewer remobilization days than the levels assumed in the postmobilization models which form the baseline case); potential improvements to training practices (or on the other hand, atrophy in the Army's experience with mobilizing RC units, leading to inefficiencies); and/or acceptance of risk by waiving certain training requirements (or on the other hand, increased training requirements for a specific major combat operation environment). Again, a move upward and to the right represents "better" performance, and the significant shift to the right with a move to shorter training times (triangle) illustrates a significant effect on demand met by the RC.

We see that changes in training time requirements have a significant effect on the percentage of demand met both by the RC and overall. Total demand met varies from around 84 percent to almost 99 per-

Figure 3.7
Effect of Changes in Postmobilization Training Time

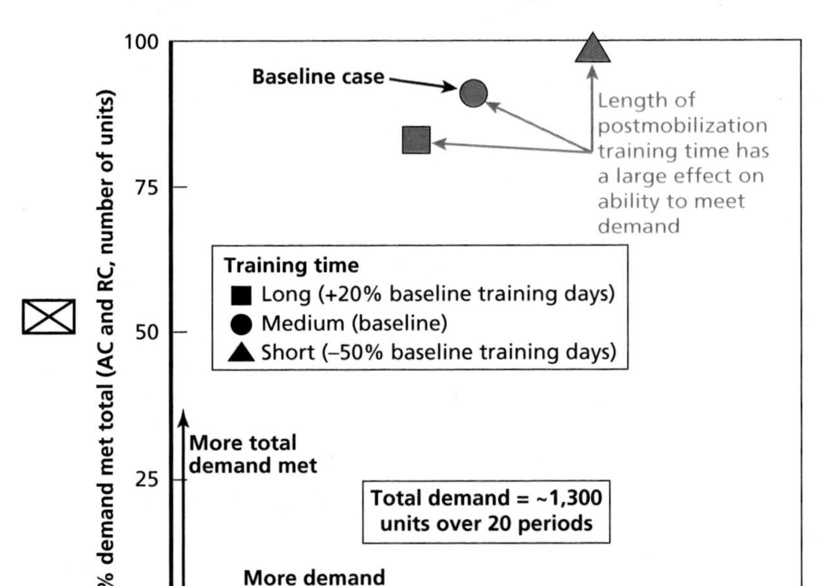

SOURCE: Outputs from RAND's high-fidelity optimization model.
RAND *RR1516-3.7*

cent when measured by number of units (83 percent to 88 percent when measured by number of soldiers) ad when moving from the longest to the shortest training time assumptions tested here. The change in percentage of demand met by the RC is even greater, with the RC meeting less than 40 percent of demand (in terms of units) under the longest training times and about 66 percent (in terms of units) under the shortest training times. The percentage of demand met by the RC in terms of number of soldiers ranges from 27 percent to 52 percent. Note that unit results represent aggregated units, and therefore assume a high degree of substitution ability between units. Actual demand met for the specific unit times requested may be significantly lower. However, our main focus here is on the effects of RC mobilization policy levers, not the absolute percentage of demand met. Also note that the shorter training time assumption represents a significant reduction from the baseline case, with the number of training days reduced by 50 percent.

Figure 3.8 displays the results shown in Figures 3.6 and 3.7 on the same plot, allowing for a direct comparison. In addition to the six cases already presented in Figures 3.6 and 3.7, Figure 3.8 adds all other combinations of training time and capacity ramp-up assumptions, yielding nine total data points (three capacity ramp-up options multiplied by three training time options). We see that more optimistic assumptions across the board lead to almost all demand being met—

Figure 3.8
Summary of Optimization Model Runs

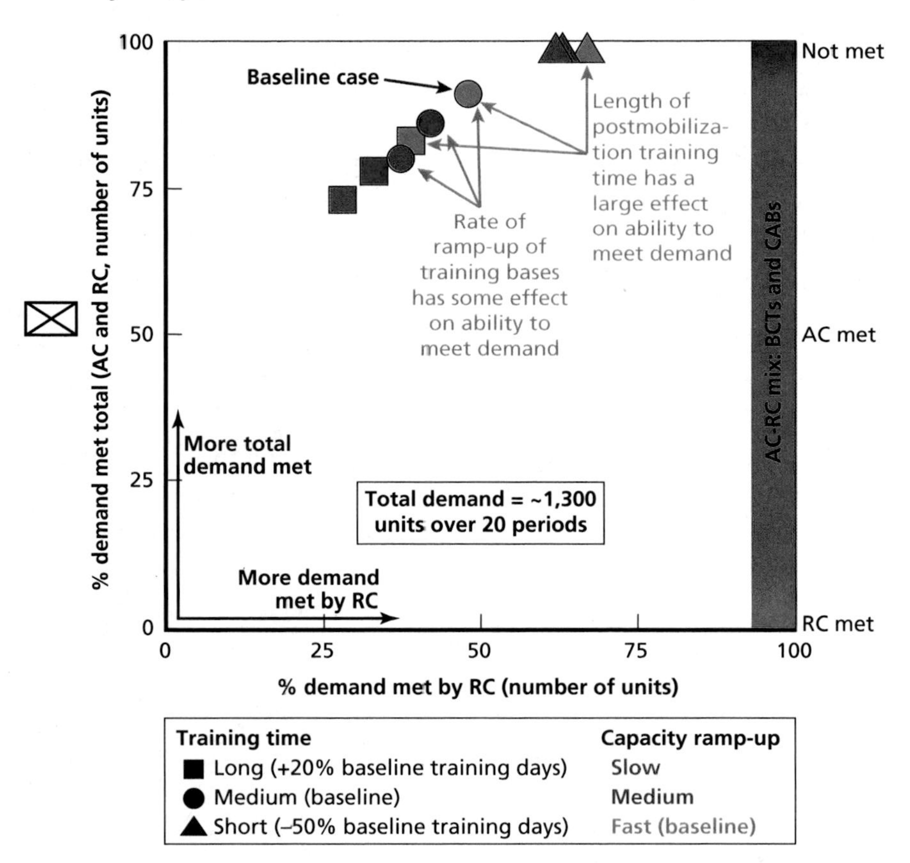

SOURCE: Outputs from RAND's high-fidelity optimization model.

RAND *RR1516-3.8*

over half by RC units. Within the boundaries of reasonable capacity ramp-up times assumed in this study, reductions in training time have a larger effect than increases in facility ramp-up speed on the ability of the RC to meet demands for a large, short-warning contingency operation. The RC's ability to meet demand shifts from a minimum of about 28 percent of units in the case with long training times and slow capacity ramp-up to a maximum of about 66 percent of units in the case with short training times and baseline (fast) ramp-up (17 percent to 52 percent when measured by number of soldiers). Overall demand met (by the AC and RC together) ranges from about 73 percent of units to almost 99 percent of units (74 percent to 97 percent when measured by number of soldiers). The model is able to achieve almost the same degree of total demand met (close to 99 percent) regardless of assumptions on facility ramp-up as long as training times are assumed to be short.

Overall, these results represent a significant increase from the demand met by the RC in the low-fidelity model; this increase is due to the fact that we are now using an optimization model to maximize the total number of RC units meeting demand by allocating capacity more efficiently.

We find that even though the RC is able to provide a high number of units, most large units (most BCTs and CABs) continue to come from the AC. This is shown by the bar to the right of the main plot in Figure 3.8, which captures the extent to which the RC can contribute CABs and BCTs to meet the demand signal. In general, yellow shading indicates CABs and BCTs supplied by the RC, green indicates those by the AC, and red indicates missed demands. Regions with gradient shading (yellow to green and green to red) represent the differences in percentages among the nine model runs included in this plot. Across the results presented in Figure 3.8, RC BCTs/CABs never meet more than about 10 percent of the total demand for BCTs and CABs, even under optimistic assumptions. In about two-thirds of the runs performed, some RC BCTs/CABs do exit the postmobilization training pipeline in time to meet the illustrative demand signal.

Differences in Results if Using Soldiers Instead of Units for Measuring Throughput

While the figures presented in this chapter focus on the number of units demanded, we can draw similar conclusions from analysis focused on the number of soldiers. Figure 3.9 parallels Figure 3.8. The gray shapes in the background of Figure 3.9 are the same nine data points presented in Figure 3.8, representing the results of all nine model runs when measured in terms of number of units. Superimposed on these results are colored shapes that show results from the same nine model runs

Figure 3.9
Soldier-Based Results

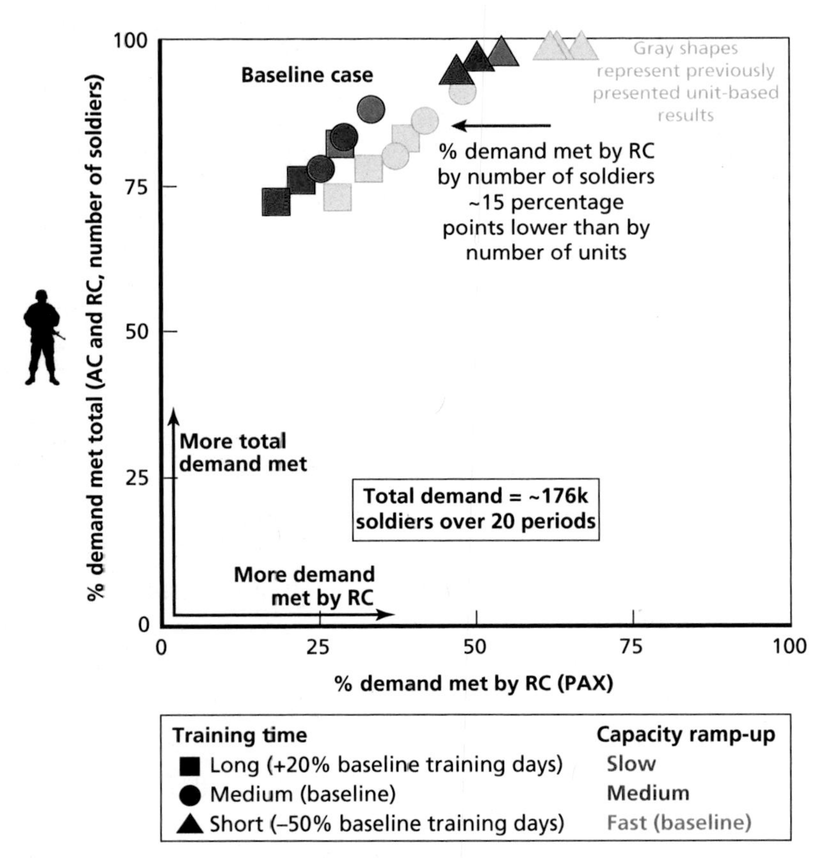

SOURCE: Outputs from RAND's high-fidelity optimization model.
RAND *RR1516-3.9*

but using a metric based on the percentage of demanded *soldiers* able to deploy. As evidenced by the fact that the colored shapes are generally slightly below but significantly to the left of the gray shapes, the shift from a unit-based to a soldier-based metric reduces the percentage met by the RC much more than it reduces the percentage of overall demand met. Across all model runs performed for this analysis, the RC's portion of demand met drops by approximately 15 percentage points when the unit of measure shifts from a unit-based to a soldier-based one. This again shows that the RC tends to provide quicker-to-train units, which tend to be smaller, under an optimized training scheme.

Throughout the rest of this report, we show results using the number of units as a measure for RC postmobilization throughput. This choice is intended to highlight the finding that larger, complex units can clog the training pipeline. While a high number of quicker-to-train, smaller units can collectively take up just as much space (and time) at an MFGI as a single CAB or BCT, an approach favoring smaller units results in a much larger number of deployable entities as requested by the COCOM commanders. This is especially relevant because variance in real-world training times and demands could potentially prevent the single large unit from being able to deploy in time to counter a real threat.

Impact of an Early Mobilization Decision

Similar to Figure 3.9, Figure 3.10 again compares nine new model runs with the baseline results originally presented in Figure 3.8. In Figure 3.10, the original results from Figure 3.8 are displayed as grayed-out shapes. Superimposed in color are the results of the same nine model runs with an earlier mobilization decision. These nine runs assume that mobilization occurs eight weeks prior to initiation of the demand signal, for example, because national leaders begin mobilization upon receiving warning of an impending conflict eight weeks before the conflict actually begins. The colored shapes are clearly far to the right of the gray shapes, indicating that the RC can meet a much larger percentage of demand when the mobilization decision is made early.

Additionally, the results for runs with slower ramp-up speed and longer training times move significantly upward when shifting to an

Figure 3.10
Effect of Earlier Mobilization Decision

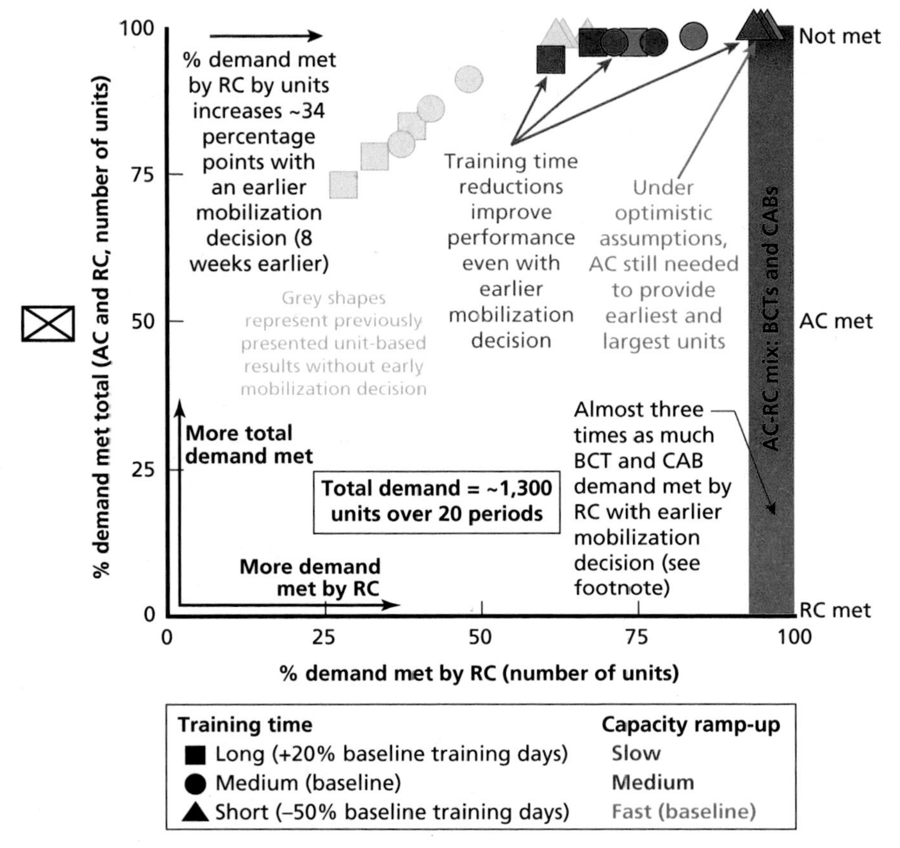

SOURCE: Outputs from RAND's high-fidelity optimization model.
NOTE: Although ~3x as much BCT and CAB demand is met by RC, the total number of RC BCTs and CABs deployed to meet TPFDD demand remains low.
RAND *RR1516-3.10*

earlier mobilization decision. This implies that the total demand met is less dependent on ramp-up and training time assumptions when the mobilization decision is made early. This is due to the increased ability of the RC to help meet demand. The demand gap becomes quite small, even under the most pessimistic training and ramp-up assumptions we used (longest training time and slowest capacity ramp-up), with almost all demand being met by either an AC or RC unit. Under the more optimistic training time and ramp-up assumptions (shorter training time and quicker capacity ramp-up) we used, the RC can supply as much as 80 percent of soldiers demanded and almost all units demanded. On average, across the assumptions we modeled, eight weeks of warning time boosts the percentage of unit demand that the RC can fill by 34 percentage points (28 percentage points when measured by number of soldiers). While this result is specific to the demand signal and assumptions used here, the general trend of a large upward shift with warning is both intuitive (more time to train yields more trained units) and generalizable to unscheduled large-scale contingency deployments.

The bar to the right of the main plot in Figure 3.10 again captures the extent to which the RC can contribute CABs and BCTs under the cases we examined. We see that, even with an early mobilization decision, the RC has difficulty generating these large maneuver units, as illustrated by the predominantly green color of the bar. The RC goes from meeting demands for one or two CABs and BCTs without any warning to meeting demands for seven or eight (about 30 percent of the BCT and CAB demand). This is limited primarily by the strict restriction on CAB and BCT training capacity; under current assumptions, the entire CAB and BCT pipeline is full when only one CAB plus two BCTs are training. In addition, BCTs and CABs have longer training times than the other unit types we modeled, meaning that even with eight weeks' warning, the first RC BCTs and CABs still are not ready on the first day of actual deployment operations, so the initial demands must be met with AC units.

Comparing Multiple Model Runs

Figure 3.11 ties our results back to the baseline runs presented in Figure 3.5 and adds the time element of the illustrative demand signal back into our analysis. The sand chart providing the backdrop of this chart is the same baseline set of results as shown in Figure 3.5. Superimposed on this graph are dotted lines indicating the effect of different assumptions on the modeled result, including options of slower ramp-up speeds and training times (most pessimistic case), shorter training times (more optimistic than baseline case), and shorter training times with an earlier mobilization decision (most optimistic case). These lines

Figure 3.11
Comparison Across Model Runs

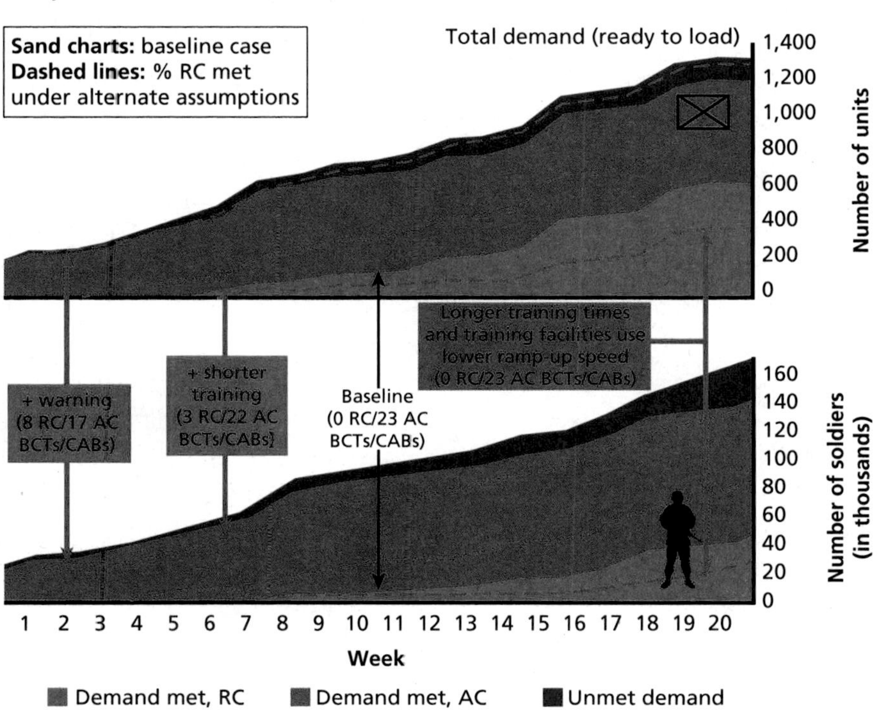

SOURCE: Outputs from RAND's high-fidelity optimization model.
RAND *RR1516-3.11*

represent the same results and model runs presented throughout this chapter, now displayed on a single graphic and using both a number of units and a number of soldiers as an output metric. These dotted lines only capture the change in contribution of RC to the demand met and not the improvement in overall demand met by both the AC and RC. In other words, the dotted lines show where the upper bound of the yellow region would be under a different set of modeled assumptions. We do not attempt to show where the upper bounds of the green and red regions would be under different assumptions. We note, however, that, as indicated in the results presented in this chapter, more optimistic assumptions allow almost all of the overall demand to be met by either the AC or RC, which would result in a sand chart with very little red area remaining.

The underlying sand chart, representing the baseline case model run, again illustrates that the RC can contribute some portion of demand for units and soldiers, especially later in the demand signal (yellow areas). The proportion of total demand met is larger when measured by number of units rather than number of soldiers because the RC more easily contributes smaller units with shorter postmobilization training times. In fact, the RC is not able to supply any CABs or BCTs in this baseline case.

In terms of the change in contribution of the RC to the overall demand signal, we begin by documenting the results of our baseline model run, which represents assumptions designed to capture capacity and training times that are similar to current plans. Because our baseline case already includes relatively optimistic assumptions about capacity ramp-up, we show results of the more pessimistic case with slow training facility capacity ramp-up and longer training times (orange dotted lines). These pessimistic assumptions about ramp-up speed and training time greatly reduce the amount of demand met by the RC in terms of both units and soldiers, although they have no effect on the RC's ability to supply large maneuver units (because none are supplied, even in our base case).

The dotted yellow lines superimposed on Figure 3.11 represent results from two cases with more optimistic assumptions. The lower set of these lines represents the baseline case with shorter training time assumptions. The higher set uses both shorter training time assumptions and the eight-week earlier mobilization decision. It is clear that both changes in assumptions provide significant marginal improvement over the baseline case, allowing the RC to contribute as many as eight BCTs and CABs and more than 95 percent of the total aggregate demand when measured by number of units (more than 80 percent of the total aggregate demand is met by the RC when measured by number of soldiers).

Overall, there is a large difference between the ability of the RC to contribute under the most pessimistic (orange dotted set of lines) and most optimistic (top yellow dotted set of lines) assumptions. Depending on decisions about timing of the mobilization decision, postmobilization training time, and facility availability and ramp-up, the RC could be expected to meet anywhere from about 28 percent to almost 96 percent of demand in terms of the number of units (about 17 percent to 83 percent in terms of number of soldiers) for a short, large deployment, such as the one examined here. The percentage of BCT and CAB demand they could help meet ranges from 0 percent under pessimistic assumptions to around 30 percent given optimistic assumptions about MFGI capacity ramp-up and postmobilization training times—which prevents clogging the training pipeline with large units that cannot be trained in time to meet demand (through the high-fidelity model's optimization feature)—and an early mobilization decision.

In short, the RC may be able to contribute anywhere from relatively little to relatively a lot, depending on the specifics circumstances of the actual operation. The set of policies, practices, and resourcing decisions related to postmobilization throughput in place at the start of the next major operation will have a significant effect on the ability of the RC to contribute both quickly (in terms of force flow onset) and to a large degree overall (in terms of percentage of the force sourced by the RC). However, current practice seems to indicate that RC contribution will be circumscribed by lack of early mobilization, lack of clarity in how mobilization sequencing priorities affect total output, and lack of mobilization capacity. In Chapter Four, we summarize the model results presented in this chapter and draw out implications and possible recommendations to policymakers and planners.

CHAPTER FOUR

Findings and Implications for RC Deployment Options

In this chapter, we summarize the detailed model outputs presented in Chapter Three and present key findings based on these modeling results. Furthermore, this chapter covers the implications of our analysis and makes several recommendations to planners for improving RC postmobilization throughput for the next major short-warning contingency deployment.

Findings

The higher-fidelity model, in conjunction with our other research and analysis, led to the findings in Figure 4.1. We expect these findings to hold in general for the case of a large, short contingency operation, in light of the caveats noted at the end of this chapter. The reader should note that they were derived specifically in the context of the demand and supply we modeled, as well as the specific assumptions we used on MFGI ramp-up capacity, RC unit postmobilization training time, and the possibility of an early mobilization decision eight weeks prior to the start of the actual conflict and the beginning of deployment operations.

In our analysis, we modeled the first 20 weeks of a large TPFDD (more than 14 BCTs with all of the command and control, combat support [CS], and combat service support [CSS] enablers; the force totaled approximately 176,000 soldiers in about 1,300 units). Under

Figure 4.1
Key Findings

Mobilization decision	• Early mobilization has the largest effect on ability to meet demand with RC inventory • Early mobilization required credible warning and/or a decision to preemptively mobilize at least some units; this will not always be feasible or politically viable • Early mobilization allows both training and facility ramp-up to begin earlier
MFGI capacity	• Within the range examined here (based on reasonable ranges specified by subject matter experts), increasing the speed at which mobilization facilities ramp up capacity has a smaller effect than early mobilization
Training time	• Policies that reduce time required for postmobilization training improve RC's ability to meet demands but may come with higher cost and/or risk (e.g., resourcing/sustaining higher levels of pre-mobilization readiness, improving training, or accepting increased risk) • RC BCTs and CABs can be trained in time to meet a sudden demand above and beyond already planned deployments only if training time is reduced and/or they are mobilized early
Unit sequencing	• If MFGI capacity remains limited, allocation of RC postmobilization training facilities must be prioritized • Large, complex RC units can sometimes clog the training pipeline • RC most efficiently provides small, quicker-to-train units • RC contribution could focus on units well suited to rapid deployment • RC will not be able to meet demands in the first few weeks, with few exceptions

SOURCE: RAND analysis of mobilization modeling results.
RAND *RR1516-4.1*

the most-pessimistic set of assumptions about capacity, mobilization timing, and training time that we modeled, we found the RC contribution could be as low as 25,000 soldiers in 400 units. Under the most-optimistic assumptions we modeled—including a mobilization decision eight weeks in advance of the actual start of the conflict—we found that the RC could contribute 140,000 soldiers and the vast majority of units, with the AC supplying only some of the largest units. However, our baseline assessment was that the RC would contribute roughly 45,000 soldiers in 600 units.[1] These are the types of inventory issues, however, that this type of analysis can inform—allowing for deliberate decisions about trading structure both within and between components.

Our analysis shows how much of the force flow to a major war could be RC, depending on several factors (e.g., timing of the decision to mobilize, postmobilization training times, MFGI capacity). As shown in Chapter Three, mobilizing earlier is the single greatest lever to increase the system's ability to produce ready RC units. If that is not possible, then decreasing the training time, prioritizing smaller, quicker-to-train units, and increasing the rate of growth of the mobilization capacity will all serve to maximize the size of the RC contribution to TPFDD, although there are additional costs and potential risks associated with these options and some only result from policy and resourcing decisions made well in advance of the start of a conflict.

Decision to Mobilize

The biggest driver is the decision to mobilize; a decision eight weeks before the TPFDD starts can double the number of RC units ready and available to meet the demands of a major war. The number of RC soldiers who deploy increases by nearly 100,000, which allows the RC to supply the vast majority of units demanded. However, it is risky to base decisions about the balance between RC premobilization and postmobilization training and/or force structure decisions on the assumption of an early mobilization decision. Therefore, this report also examines more feasible policy levers, although their effects are more modest.

MFGI Capacity

Changes to MFGI capacity ramp-up, at least within the bounds deemed reasonable by subject-matter experts consulted for this study, have a relatively lesser effect on RC postmobilization training through-

[1] It is important to note that regardless of the optimism of our assumptions, there was a low number of demands not met by either AC or RC forces. These represent a total Army inventory issue, not a readiness issue. Addressing inventory shortfalls is outside of the scope of this analysis.

put as measured by number of units or number of soldiers. Consistent with expert opinion, we assume that the number of MFGI facilities available for RC postmobilization training (and their total capacity) is fixed. We vary the rate at which this capacity becomes available within relatively narrow bounds that could be feasible without a significant change in investment. Across our runs, the number of days required to achieve 50 percent of total available MFGI capacity varies from ten weeks to 14 weeks.

Training Time and Unit Sequencing

An early mobilization decision could be unrealistic, and MFGI capacity and ramp-up are relatively inflexible without a significant increase in investment. In contrast, reduced postmobilization training time and well-planned sequencing of units through the pipeline could provide a more feasible avenue for increasing postmobilization throughput. As discussed earlier, an effort to shorten postmobilization training has a variety of resource and risk implications that we do not explore deeply.[2]

Even without actual reductions in training times for individual units, however, training times have a greater effect; unit type (and related size and complexity) is inherently linked to training time, with larger more complex units requiring longer stays at MFGI facilities. For this reason, the mix of RC units put through the mobilization pipeline has a big effect on how many trained RC units or soldiers can be produced in time. Pushing through BCTs and CABs early in the mobilization process severely limits the ability to get smaller RC units out in time because BCTs and CABs both monopolize available training capacity and take a relatively long time to train. These large, complex units are effectively delaying smaller RC units from processing through the system. Also, because BCTs and CABs take so long to train, few make it out of the pipeline in time. This implies a clear trade between mobilizing BCT and CABs versus other RC capabilities.

---

[2] Key areas of concern are increased cost, increased demands on RC soldiers, and increased levels of risk. This problem set is highly developed in other research described in the introductory paragraphs and footnotes.

Arbitrating this trade can involve both force structure and operational considerations.

Absent a compelling operational need for RC BCTs and CABs in the late stages of TPFDD demand (around the 150- to 180-day mark), this suggests that the RC should initially focus its contributions on relatively small units that are quicker to train and mobilize and can affect the early phases of a campaign.[3] Waiting until MFGI capacity has expanded to accommodate both CABs and BCTs (at dedicated facilities) *and* additional smaller units (at other facilities) prevents clogging the training pipeline. This could be done in several ways. A first step could be to make prioritizing small, quicker-to-train units an explicit goal in mobilization planning. Beyond that, resources could be allocated to raise the peace-time readiness of such units. Ultimately, adjustments to the force structure between the AC and RC components could be explored (such adjustments are beyond the scope of this study). However, this should not be done solely on the basis of this work; a much broader set of factors needs to be considered, including the demand for RC BCTs and CABs in the middle and late phases of a campaign, for homeland defense, or even early campaign combat operations where they do not need to train above battalion level.

Implications

The focus of this study is an analysis of how to increase the ratio of RC units to AC units deploying in the early weeks of a major crisis by changing how the Army executes mobilization and contingency planning. The study is not a normative assessment about whether to

---

[3] The military has a formal definition of phases that represent a natural progression and subdivision of a campaign or operation. Joint planning is based on a six phase construct: shape (Phase 0), deter (Phase 1), seize the initiative (Phase 2), dominate (Phase 3), stabilize (Phase 4), and enable civil authority (Phase 5). The dominant phase (Phase 3) focuses on breaking the enemy's will for organized resistance or, in noncombat situations, control of the operational environment (see Joint Publication 5-0, *Joint Operational Planning,* Washington, D.C.: Joint Chiefs of Staff, August 11, 2011). Training tasks required to be ready for operations in Phase 3 can be very different from those required for Phase 4 or Phase 5.

change the ratio, nor is it a comprehensive enough study to suggest force structure changes that might be inferred (rightly or wrongly) from application of the results. However, the study clearly suggests that a bias toward focusing on the training and contingency sourcing of more readily deployable units in the RC will greatly increase the RC's ability to contribute in early phases of a future deployment. This, in turn, implies that the RC's CS and CSS structure is more important to achieving that change in ratio than is the RC's combat structure because of how the complexity and size of the RC's combat forces affect the overall mobilization process. And, this, in turn, suggests that contingency sourcing of war plans should be correspondingly focused on early (and appropriate) use of the RC's CS and CSS structure rather than early use of RC combat forces, when AC structure is not available or when there is a desire to include RC forces in the deployment.[4] Of course, a look at the middle and late phases (as the mission profile begins to change from rapid decisive operations and into transition and stability operations) may provide different insights and structure recommendations. These phases, however, occur when rapid mobilization and deployment have ceased to become a key operational limiting factor. They may, however, represent an area of comparative advantage for other parts of the RC force structure.

Assuming early mobilization decisions do *not* occur, it is important to note that even small RC units will have trouble meeting demands in the first 30 to 60 days. This means that a "full set" of enablers (in either the AC or the RC) for the AC forces that are needed for that period should be kept at especially high readiness and be prepared to deploy on short notice. This suggests that a narrow and definable section of the RC should be kept at high readiness, which is likely more feasible than sustaining high readiness across a large cross-section of the RC. This narrow and definable section would be based on analysis of AC

---

[4] We realize that the Army National Guard has combat units and the Army Reserve does not, so much of this discussion may be much more relevant to Army Reserve force structure—always and continually caveated by the need to incorporate both homeland defense and defense support of civil authorities requirements, which we have not considered in our analysis.

lack of depth or availability in that capability area—and may change as AC deployments occur or as AC force structure changes.

Overall, our findings have the following implications for RC policies, investments in RC readiness, and the balance of force structure across the components of the Total Force:

- If planners are willing to assume that selective early mobilization is possible, then the RC can make a more substantial contribution in a major contingency surge. However, this is a high-risk assumption: If national leaders do not authorize early mobilization, then it is unlikely that the necessary forces will be able to deploy on time. Our analysis indicates that the other levers available to accelerate RC deployment are unlikely to be able to compensate, even if funding is unconstrained once the war starts. Identifying the RC units necessary to support the TPFDD demand and selectively mobilizing those units and the MFGI facilities can provide an option between full mobilization and no mobilization. However, it requires significant planning and a level of coordination across Army components, geographic combatant commands, and TRANSCOM that does not exist today. Moreover, early mobilization, whether in whole or part, requires advance warning of impending conflict. The United States cannot always count on such a warning, meaning that using advance warning as a key planning assumption involves a layer of risk independent of whether national leaders would actually choose to preemptively mobilize; they may simply not have the choice in some cases.
- Regardless of when the mobilization decision is made, it is unlikely that more than a few RC BCTs and CABs can be mobilized and made ready in time to meet TPFDD demands. This is a function both of the relatively long postmobilization training times for such units, as well as the current limits on MFGI capacity that permit only one CAB and one or two BCTs to go through postmobilization training at one time. From this perspective, our findings support the current bias toward BCTs and CABs in the AC force structure and toward smaller units that can be trained more quickly in the RC force structure. However, many

other considerations influence the mix of force structure across the Total Force—such as homeland defense, stability operations, and rotational support to a long war—that argue for maintaining a number of BCTs and CABs in the RC. But, from the narrow perspective of "what can the RC get out the door in time for a major war," this analysis points quite clearly toward biasing the RC toward smaller units with relatively shorted postmobilization training times.

- While increases in postmobilization training capacity, as well as the ability to ramp up such capacity quickly, could also improve RC postmobilization training throughput, our analysis suggests that the type of small increases that would be reasonable (given the existing capacity and level of investment) would have a much lesser effect than the type of sequencing and readiness decisions mentioned earlier.
- Assuming there is no early mobilization, even smaller RC units (with some limited exceptions) are likely to have trouble meeting demands in the first 30 to 60 days of TPFDD execution. This suggests that the set of enablers for deploying AC forces that are needed in this period should either be maintained in the AC or kept in the RC but maintained at especially high readiness and be prepared to deploy on 30 days' notice. Further study would be needed to ensure that maintaining such high levels of RC readiness in peacetime is feasible.

Recommendations

The discussion above suggests that optimizing usage of the RC in support of a major conflict would mean the following:

- Focus on deploying smaller, quicker-to-train RC units in the earlier periods of a conflict, and defer the larger, more-complex RC formations to later stages of major operations and transition or stabilization operations. This will require appropriate planning by

Army components, combatant commands, and TRANSCOM, as well as perhaps some adjustments in force mix across the AC and RC. Focusing on this type of planning effort may also provide for faster and more practiced exercise of the mobilization process and for identification of possible changes other than those discussed in this report.

- Focus investments on maintaining readiness in the types of RC units that must or should deploy early. Later-deploying RC units will have sufficient time to train. Additionally, if these units deploy after the initial force flow, their training can focus on transition and sustainment operations, not major combat operations. Investments in early-deploying units may include not only training dollars but also training seat allocations, overmanning, and other actions that improve units' general readiness when they must mobilize with no notice. Again, further study would be needed to ensure that maintaining such high levels of RC readiness in peacetime is feasible.
- Consider re-creating the WARTRACE and CAPSTONE-like process of matching specific units (at the unit identification code level) to the TPFDD demands to better focus peacetime and post-mobilization training.[5] This may also allow for a more nuanced

[5] CAPSTONE and WARTRACE were programs developed and implemented in the 1970s and 1980s to improve RC training and better focus RC units on their likely wartime missions. See John O. Marsh Jr., *Subject: Problems in Implementing the Army's CAPSTONE Program to Provide All Reserve Components with a Wartime Mission,* Secretary of the Army memorandum to the Army Inspector General, Washington, D.C., GAO/FPCD-82-59, September 22, 1982, which states:

> The Army's CAPSTONE program is designed to align all Army reserve component units—Army National Guard and Army Reserves—under gaining-commands (those commands which will employ Reserve units in wartime) and provide units with detailed information concerning their wartime mission. This information is to be used to improve wartime planning and ongoing training for Reserve component units.

and Army Regulation 11-30, *Army WARTRACE Program,* Washington, D.C.: Headquarters, Department of the Army, July 28, 1995, which states:

> The WARTRACE Program dates back to 1973 when the Affiliation Program was approved. The intent of Affiliation was to improve the training and readiness of RC combat battalions and brigades by associating them with AA units. Under this program,

description of what makes a unit "ready enough" to deploy, and therefore save some training time postmobilization and allow for greater throughput. This approach could include early identification by COCOMs of critical theater-entry training and readiness requirements, enabling better focus of both AC and RC unit training plans and concomitant shortening of postalert and postmobilization training timelines.

Caveats

There are several important caveats to the recommendations:

- Cost is not a factor we considered directly, although there are obviously different—potentially significantly different—costs associated with the options identified.
- This analysis is about contingency response for major overseas wartime or conflict demands. It does not account for domestic missions (either defense support to civil authorities or homeland defense), nor does it consider non-conflict deployments (e.g., humanitarian assistance or disaster response) or known, scheduled deployments (Global Force Management-sourced and approved). Each of these represents ripe grounds for RC usage that influences RC policies, RC readiness, and balance of force structure across the Total Force.[6]

> AA divisions formed training relationships with [Army National Guard] and [Army Reserve] units and worked with these units during both AT and IDT. In 1976, CS and CSS units were added to the program. In 1978, two [Army National Guard] divisions were linked with two AA divisions under the Division Partnership Program to increase the readiness of [Army National Guard] divisions.

For a holistic look at the two practices, see also Dennis P. Chapman, *Planning for Employment of the Reserve Components: Army Practice, Past and Present,* Arlington, Va.: The Institute of Land Warfare, September 2008.

[6] For a more detailed look at some of the issue affecting AC/RC force mix decisions, see Chris Marie Briand, *Operational Army Reserve Sustainability Fact or Fiction?,* Carlisle, Pa.: U.S. Army War College, 2016.

- We held constant the number of primary MFGIs and held as given the maximum capacity of each of those MFGIs. We did not examine the cost effectiveness of funding, in peacetime, a significantly larger "hot base" for mobilization.

APPENDIX A

Shipping Considerations for Limiting Deployment Speed

In large contingencies, sealift delivers the vast majority of vehicles and bulk supplies (including fuel and ammunition). Real-world physics limits the speed of sealift vessels and therefore the rate at which the deployment force flow can unfold. This appendix provides evidence supporting the finding that, under optimistic assumptions about sealift availability, TPFDD are already developed as transportation-feasible plans and thus do not meaningfully underestimate the amount of time available for RC units to get ready to deploy. While baseline transportation assumptions (using only Department of Defense-owned assets) may not provide enough shipping capacity to meet TPFDD timelines, sufficient lift could be made available through the use of contracted commercial shipping. This claim justifies our focus on postmobilization training throughput as the main limiting factor in determining how much of the force flow can be met with ready RC units. Limitations on shipping speed and throughput vary based on geography, the location of prepositioned stocks, and the availability of shipping capacity. Department of Defense-owned shipping capacity is the most limiting factor under today's baseline assumptions, but it can be supplemented through use of new and existing contracts with commercial carriers.

To assess factors that affect sealift capacity, we studied delivery capacity responding to a representative demand signal similar in size and force mix to the TPFDD-based signal used throughout this report. We used outputs from TRANSCOM Surface Deployment and Distribution Command (SDDC) shipping models to examine the effect that various increased capacity levers have on the ability to make the specific set of required deliveries for this demand signal. The illustrative demand signal used by TRANSCOM SDDC includes sufficient planning details to simulate the nuances of an actual deployment, including readiness issues (especially early in the delivery process) and capacity use issues (especially late in the delivery process). Figure A.1 presents an illustrative delivery requirement (solid black) and the modeled deliveries achieved using a conservative baseline set of shipping capacity assumptions based on only Department of Defense-owned assets (dotted black). The colored lines represent marginal improvements—that is, delivery of the required short tons sooner—on this baseline achieved by adjusting various policy measures, which we examine in more detail later in this appendix. Figure A.1 shows that assumptions about the number of available ships—either maritime prepositioning ship squadron (MPSRON) vessels that have already completed their initial delivery of Marine Corps equipment, commercial ships available through the VISA program, or other roll-on/roll-off (RORO) vessels—have a greater effect on the ability to deliver the requirement.[1]

VISA is a government-commercial partnership program administered by TRANSCOM. This program ensures the government's access to commercial shipping in case of emergency deployment and sustainment needs.[2] Under the program, commercial shipping companies register their fleet capacity to receive preferential government contracts and other benefits. This program provides up to 305,000 short tons of capacity on 361 vessels owned by 55 carriers. These ships include bulk carriers, container ships, heavy lift ships, ROROs, barges, tugboats, and other specialty ships. While the program has never been activated, it improves government awareness of shipping capability and relationships with commercial shippers and allows for use of contract-based

[1] For the analysis in this section, we collaborated heavily with SDDC-TEA. Much of the work is based on outputs from their AMP model in response to data and questions we provided. The analysis of the results is ours, however.

[2] See Maritime Administration, "Voluntary Intermodal Sealift Agreement (VISA)," undated-b.

commercial shipping when necessary. Activation of VISA can occur in stages; VISA III activation can generate over three times as much capacity as VISA II because it taps into non-Jones Act fleet capacity and activates all carriers participating in the Maritime Security Program.[3] Jones Act ships are those that are built in the United States, owned and crewed by Americans, and fly the American flag. Expanding to the non-Jones Act fleet would mean allowing non-U.S.–flagged and crewed vessels owned by the carriers involved in the Maritime Security Program to carry military supplies and equipment. Beyond the already preregistered VISA vessels, the Department of Defense could further increase capacity though one-time contracts.

In Figure A.1, each line represents a marginal improvement over the baseline. For example, the bright blue VISA II line indicates performance of the baseline plus VISA II activation. The green line labeled "+30 ROROs and 2x rail" represents performance of the baseline with an additional 30 ROROs and with double the railcar capacity. While none of the runs presented here fully close the gap between the baseline and requirement, the addition of 30 ROROs plus additional railcars results in a delivery curve that is only a few days short of the requirement. While a small number of days can be important in a real-world operation, this illustrative analysis serves to illustrate that the addition of shipping and rail capacity can close much of the gap between the baseline and illustrative requirement.[4] Transportation is likely to be a constraining factor in any large-scale deployment, but the analysis presented in this appendix illustrates that, at least under optimistic assumptions and a large investment in shipping capacity at the time

---

[3] Jones Act fleet refers to ships governed by the Merchant Marine Act of 1920 (U.S. Code, Title 46, Chapter 24, Merchant Marine Act, 1920), also known as the Jones Act. This federal statute governs the promotion and maintenance of the United States Merchant Marine. For more on the Maritime Security Program, see Maritime Administration, "Maritime Security Program (MSP), undated-a.

[4] Because this is all illustrative data, we did not try to calculate the exact numbers of ships or rail cars required to match the delivery timeline. The clear implication is that increased access to shipping and rail cars—through purchase or lease—can allow timely closure of even very large forces. This truncated analysis allowed us to focus our efforts on the role of mobilization and postmobilization training.

Figure A.1
Sealift Constraints and Factors Affecting Capacity

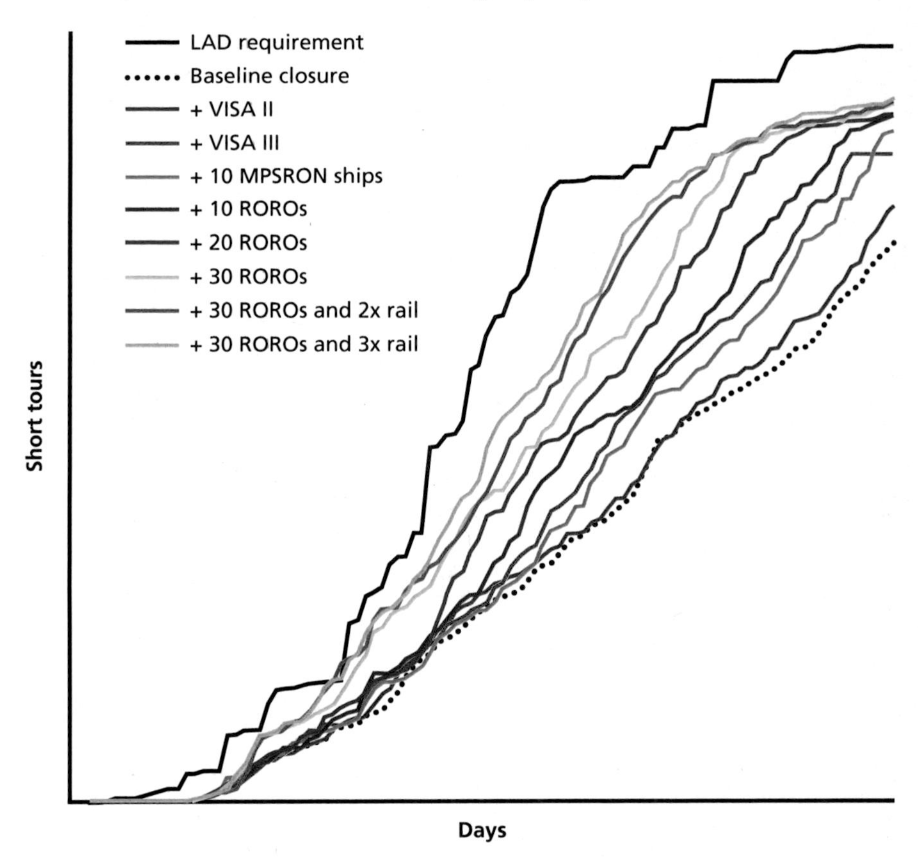

SOURCE: Based on BBN Technologies; see SDDC-TEA, "Analysis of Mobility Platform (AMP)," U.S. Army, undated.
NOTE: *2x* and *3x* rail refer to doubling or tripling the number of rail cars available to move equipment to ports of debarkation. + = implies access to that number of additional ships indicated for transport of equipment over the baseline.
RAND *RR1516-A.1*

of conflict, it could be possible to come close to delivering even an extremely large force package on an aggressive timeline.

Examining factors such as the time it takes to make initial deliveries, the time it takes to deliver 80 percent of the total requirement, and the fastest delivery rate seen during the peak of the demand signal allows us to quantify the effect of different transportation assumptions and to evaluate the sufficiency of today's transportation capacity relative to the demand signal. We examine the effect of the availability of commercial shipping either through the activation of various stages of the VISA program or through other contracting for additional RORO vessels, the use of MPSRON ships, access to increased rail transport capacity, and changes in preferred ship utilization rates. Additional rail assets and changes in ship utilization rates have a relatively lesser effect. Determining whether additional rail assets are readily available is outside of the scope of this study. Adding ships, whether they come from the Marine Corps through MPSRON, or commercial contracts through VISA or other agreements, has a much greater effect. Access to additional ships increases capacity *in each delivery cycle*. The size of the effect from these shipping policy changes primarily relates to the number and total capacity of ships that become available, with VISA III activation (the highest stage of VISA program usage) providing the largest number of ships of the options explored here.[5]

In an emergency deployment situation, the first commercial ship could generally be available for pickup in the continental United States about one week after activation, having offloaded its commercial cargo first. With the use of VISA and other contract-based commercially available shipping (represented in Figure A.1 by the addition of ROROs), much of the gap between the illustrative requirement and baseline delivery can be closed.[6] The degree to which shipping capacity limits deployment speed and execution of a war plan is highly dependent on the specific circumstances of a given conflict. Differences in geography, warning, contingency size, and levels of port access and anti-access threats will determine to what extent the factors explored in this report affect the outcome of a contingency. However, the sheer number of ships available in the commercial sector is large. This large

---

[5] As more ships are made available, the throughput bottleneck will eventually shift away from shipping capacity and toward other issues, such as readiness of individual units and port infrastructure. However, we find that up to 40 additional RORO ships, which could come from VISA or other commercial shipping capacity, can be efficiently employed (meaning that their use brings about an improvement in delivery day and flow) using baseline assumptions about unit readiness, port infrastructure, and rail assets.

[6] This report does not investigate the political feasibility of VISA activation.

pool can be leveraged, at least theoretically, during times of conflict. While transportation continues to be a significant constraint on deployment speed, the analysis in this appendix shows that, given optimistic—but not unreasonable—assumptions and a large investment in shipping capacity at the time of conflict, it would be possible to come very close to delivering even an extremely large force package on a very aggressive timeline. Therefore, examining the Army RC's ability to meet such an aggressive timeline is necessary.

APPENDIX B

Mixed-Integer Optimization Model for RC Mobilization

This appendix provides the notation and formal mixed-integer program that is used to solve for the optimal RC mobilization under various scenarios and cases. The optimization model was developed in the General Algebraic Modeling System (GAMS),[1] a modeling system for mathematical optimization, and solved using the mixed-integer programming solver in CPLEX.[2] All model runs were completed until an optimality gap of 0.2 percent or less was achieved.

Generically, the mixed-integer optimization model we employ is the solution to a constrained optimization with a linear objective, linear constraints, and continuous, integer and binary variables (the sets of variables x, y, and z, respectively). Thus, it can be written as:

[1] General Algebraic Modeling System (GAMS), Version 23.3, homepage, undated, accessed January 18, 2016.

[2] IBM, "CPLEX Optimizer: High-Performance Mathematical Programming Solver for Linear Programming, Mixed Integer Programming, and Quadratic Programming," undated.

$$
\begin{aligned}
\min\ & c_1^T x + c_2^T y + c_3^T z \\
\text{s.t.}\ & A_1 x \leq b_1 \\
& A_2 y \leq b_2 \\
& A_3 z \leq b_3 \\
& x \in R_+^{n_1} \\
& y \in Z_+^{n_2} \\
& z \in \{0,1\}^{n_3}
\end{aligned}
\tag{1}
$$

As the y and z decision variables are discrete, (1) is—like all mixed-integer programs of this type—an NP-hard optimization problem (that is, any algorithm to solve the problem that runs in polynomial time will be nondeterministic),[3] as the feasible region is nonconvex. Because this general formulation does not provide insight into the particular constraints in the RC mobilization optimization model, we provide a detailed version of mixed-integer program in the next section.

RC Optimization Model

We introduce the notation for the parameters and decision variables before providing the specific RC optimization model. We then provide a brief explanation of the model's major components (objective function and constraints). Many of the parameter values depend on the scenario modeled and can be found in Chapter Three (e.g., Figure 3.2 shows the values for many of the capacity parameters).

Parameters

We let:

$t = 1,\ldots T$ be the number of time periods

[3] George L. Nemhauser and Laurence A. Wolsey, Integer and Combinatorial Optimization, New York: Wiley & Sons, 1988.

C_t	be the training capacity at t for all locations, except Fort Hood, Fort Bliss, and Camp Shelby
$CB_t, \hat{CB}_t$	be the training capacity at t for Fort Bliss for BCT and non-BCT units, respectively
$CS_t, \hat{CS}_t$	be the training capacity at t for Camp Shelby for BCT and non-BCT units, respectively
$CH_t, \hat{CH}_t$	be the training capacity at t for Fort Hood for CAB and non-CAB units, respectively
N_1	be the set of all BCT pacing units
N_2	be the set of all CAB pacing units
N_3	be the set of non-BCT and non-CAB units
$i \in N$	be the set of all units (i.e., $N = N_1 \cup N_2 \cup N_3$)
u_i, u_i^{AC}	be RC and AC, respectively, supply of unit type i
s_i	be the size of unit type i
w_i	be required training window (number of weeks) for unit type i
$d_{i,t}$	be the demand of unit type i at t
rBN_i	be the ratio of maneuver units to other units in BCT or CAB ($i \in \{N_2, N_3\}$)
p_i	be the penalty for failing to meet a unit demand of type i with an RC
wt_{RC}	be the weight for the penalty for failing to meet a unit demand of type i with an RC
wt_{AC}	be the weight for the reward for meeting a missed demand of type i with an AC
M	be an arbitrarily large, positive number

Decision Variables

We have the following integer, continuous and binary variables:

$x_{i,t} \in Z_+^N$	the number of RC units of type i that start training at t
$xB_{i,t} \in Z_+^{N_1}$	the number of RC BCT pacing units of type i that start training at Fort Bliss at t

$xS_{i,t} \in Z_{+}^{N_1}$	the number of RC BCT pacing units of type *i* that start training at Camp Shelby at *t*
$x_{i,t}^{AC} \in Z_{+}^{N}$	the number of AC units of type *i* that are used to meet RC-missed demand at *t*
$y_i \in Z_{+}^{N_1 \cup N_2}$	indicator to ensure proper CAB and BCT training ratios
$f_{i,t} \in R_{+}^{N}$	the number of units of type *i* demanded at *t* that were RC-missed
zB_t, zS_t	binary variables for whether any BCT training occurs at *t* at Fort Bliss (or Camp Shelby)
zH_t	binary variable for whether any CAB training occurs at *t* at Fort Hood

Mixed-Integer Program

We then can formulate and solve the following optimization problem:

$$\min \sum_{i,t} wt_{RC} p_i f_{i,t} - \sum_{i,t} wt_{AC} p_i x_{i,t}^{AC} \tag{2}$$

$$\text{s.t. } d_{i,t} \le \sum_{\tau=1}^{t-w_i} x_{i,\tau} - \sum_{\tau=1}^{t-1} (d_{i,\tau} - f_{i,\tau}) + f_{i,t} \forall i, t = w_i + 1, \ldots, \mathrm{T} \tag{3}$$

$$d_{i,t} = f_{i,t} \quad \forall i, t = 1, \ldots, w_i \tag{4}$$

$$d_{i,t} \ge f_{i,t} \quad \forall i, t = 1, \ldots, w_i \tag{5}$$

$$\sum_{i,t} \left(x_{i,t} + f_{i,t} \right) \ge \sum_{i,t} d_{i,t} \tag{6}$$

$$\sum_{i \in N_3} \sum_{\tau = t - w_{i+1}}^{t} s_i x_{i,\tau} \le C_t + C\hat{B}_t \left(1 - zB_t\right) + C\hat{S}_t \left(1 - zS_t\right) + C\hat{H}_t \left(1 - zH_t\right) \ \forall t \tag{7}$$

$$CB_t - \sum_{i=1}^{N_1} \sum_{\tau=1-w_i+1}^{t} s_i xB_{i,\tau} \ge 0 \ \forall t \quad CS_t - \sum_{i=1}^{N_1} \sum_{\tau=1-w_i+1}^{t} s_i xS_{i,\tau} \ge 0 \ \forall t \tag{8}$$

$$CH_t - \sum_{i \in N_2} \sum_{\tau=1-w_{i+1}}^{t} s_i x_{i,\tau} \geq 0 \qquad \forall t \tag{9}$$

$$\sum_{i=1}^{N_1} \sum_{\tau=t-w_{i+1}}^{t} xB_{i,\tau} \leq MzB_t \ \forall t \qquad \sum_{i=1}^{N_1} \sum_{\tau=t-w_{i+1}}^{t} xS_{i,\tau} \leq MzS_t \ \forall t \tag{10}$$

$$\sum_{i \in N_2} \sum_{\tau=t-w_i+1}^{t} x_{i,\tau} \leq MzH_t \qquad \forall t \tag{11}$$

$$x_{i,t} = xB_{i,t} + xS_{i,t} \qquad \forall i = 1,..,N_1, t \tag{12}$$

$$\sum_{i=1}^{N_1} \sum_{\tau=1-w_1+1}^{t} xB_{i,\tau} \leq 3 \ \forall t \qquad \sum_{i=1}^{N_1} \sum_{\tau=t-w_1+1}^{t} xS_{i,\tau} \leq 3 \ \forall t \tag{13}$$

$$\sum_{i \in N_2} \sum_{\tau=t-w_i+1}^{t} x_{i,\tau} \leq 2 \qquad \forall t \tag{14}$$

$$\sum_{t=1}^{T-w_i} x_{i,t} \leq rBN_i y_i \ \ i \in N_1 \qquad \sum_{t=1}^{T-w_i} x_{i,t} \leq rBN_i y_i \ \ i \in N_2 \tag{15}$$

$$\sum_t x_{i,t} \leq u_i \qquad \forall i \tag{16}$$

$$\sum_t x_{i,t}^{AC} \leq u_i^{AC} \qquad \forall i \tag{17}$$

$$x_{i,t}^{AC} \leq f_{i,t} \qquad \forall i,t \tag{18}$$

$$x_{i,t}, x_{i,t}^{AC}, y^i \in Z_+, zB_t, zS_t, zH_t \in \{0,1\}, f_{i,t} \geq 0 \tag{19}$$

The objective function (2) has two components: minimizing the amount of demand that the RC units cannot fill and maximizing the amount that is met by the AC units. If p_i is the same value for all i (i.e., $p_i = p_j \ \forall \ i \neq j$), then meeting the demand with RC preferentially over AC will occur if $wt_{RC} > wt_{AC}$. Because there is a negative penalty (i.e., a "reward") for using AC units (i.e., $x_{i,t}^{AC}$), demand that is unmet by the RC, $f_{i,t}$, will then be satisfied by the AC if there is available AC supply. As a result, the objective function ultimately minimizes the total unmet demand but uses RC units preferentially because $wt_{RC} > wt_{AC}$.

The following provides a brief summary of constraints (3)–(18).

- (3): Demand of unit type *i* at time *t* that cannot be met by RC is classified as RC-missed
- (4): Demand of unit type *i* that occurs prior to the minimum training time cannot be met by RC
- (5): In combination with constraint (3), this constraint ensures that demand that cannot be met at *t* is classified as RC-missed at *t* (as opposed to t', where $t' \neq t$)
- (6): No excess training beyond the demand is possible
- (7): Non-BCT and non-CAB training cannot exceed the capacity
- (8): BCT training at Fort Bliss or Camp Shelby cannot exceed capacity
- (9): CAB training at Fort Hood cannot exceed capacity
- (10): Limit of capacity at Fort Bliss or Camp Shelby if there is BCT training at *t*
- (11): Limit of capacity at Fort Hood if there is CAB training at *t*
- (12): Total BCT training is the sum of BCT training at Fort Bliss and Camp Shelby
- (13): No more than three BCT pacing units can be training at time *t* at Fort Bliss or Camp Shelby
- (14): No more than two CAB pacing units can be training at time *t* at Fort Hood
- (15): Training of BCT/CAB pacing units is at the appropriate ratio to other units
- (16): Training RC unit type *i* must be no greater than its supply
- (17): Use of AC unit type *i* cannot be greater than its supply
- (18): Use of AC unit type *i* is limited to the amount of RC-missed

APPENDIX C

Robustness of the AC Readiness Assumption

Figure C.1 plots the percentage of RC demand met (blue) and total demand met (green) across runs of the optimization model that vary the 50-percent assumption on AC readiness. In the baseline models presented in this study, we assume that 50 percent of AC inventory of each unit type is ready enough to meet their required demand date. We see that the percentage of demand met by the RC is unaffected by changes in this 50-percent assumption, indicating that our results on RC usage are not sensitive to this assumption. The percentage of total demand met increases as AC availability increases because additional ready AC units are available to meet these demands. In other words, as one increases the AC inventory (the x-axis), the total demand satisfied increases (green curves), but the RC contribution to that demand satisfaction (blue lines) is unchanged. The AC readiness assumption does not affect our analysis because the analysis gives first priority to RC when filling demands. Only if RC is not available are AC units used. Therefore, AC readiness affects the amount of unmet demand but not the amount of RC used to meet demand. Nevertheless, it is the case that a different AC/RC force mix would affect the ability to meet the demand, as the model would potentially prioritize RC units differently.

Figure C.1
Robustness of the AC Readiness Assumption

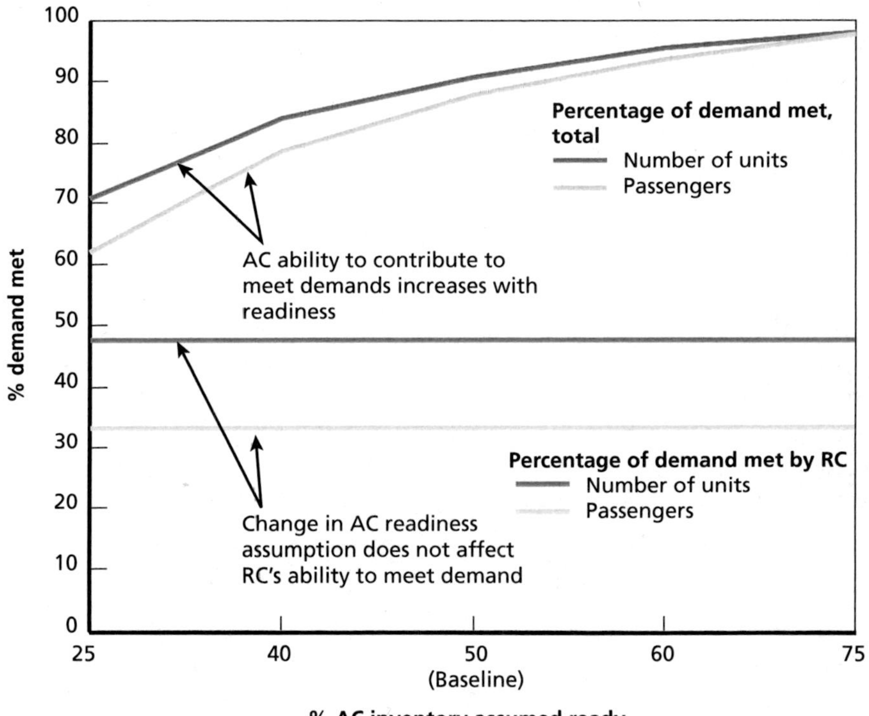

SOURCE: Outputs from RAND's Mixed-Integer Optimization Model for RC Mobilization.

RAND *RR1516-C.1*

Abbreviations

ABCT	armored brigade combat team
AC	active component
AOR	area of operations
BCT	brigade combat team
CAB	combat aviation brigade
COCOM	Combatant Command
CONPLAN	concept plan
CS	combat support
CSS	combat service support
FORSCOM	U.S. Army Forces Command
GAO	U.S. Government Accountability Office
HLD	homeland defense
HQDA	Headquarters, Department of the Army
JBLM	Joint Base Lewis-McChord
JBMDL	Joint Base McGuire-Dix-Lakehurst
JFC	Joint Force Commander
JOPES	Joint Planning and Execution System

LAD	latest arrival date
MEB HQ	Maneuver Enhancement Brigade Headquarters
MFGI	Mobilization Force Generation Installation
MPSRON	maritime prepositioning ships squadron
MTOE	Modified Table of Organization and Equipment
OASD(RA)	Office of the Assistant Secretary of Defense for Reserve Affairs
OSD	Office of the Secretary of Defense
RC	reserve component
RORO	roll-on/roll-off
SDDC	Surface Deployment and Distribution Command
SDDC-TEA	Surface Deployment and Distribution Command Transportation Engineering Agency
SRC	standard requirements code
TPFDD	time-phased force and deployment data
TRANSCOM	United States Transportation Command
VISA	Voluntary Intermodal Sealift Agreement

References

Army Regulation 11-30, *Army WARTRACE Program,* Washington, D.C., Headquarters, Department of the Army, July 28, 1995.

Army Regulation 220-1, *Army Unit Status Reporting and Force Registration—Consolidated Policies,* Washington, D.C.: Headquarters, Department of the Army, April 15, 2010.

Bates, James C., "JOPES and Joint Force Deployments," PB700-04-3, *Army Logistician*, Vol. 36, No. 3, May–June 2004.

Briand, Chris Marie, *Operational Army Reserve Sustainability Fact or Fiction?* Carlisle, Pa.: U.S. Army War College, 2016.

Chairman of the Joint Chiefs of Staff, *Joint Operation Planning and Execution System (JOPES), Vol. 1, (Planning Policies and Procedures)*, Chairman of the Joint Chiefs of Staff Manual 3122.01, Washington D.C.: July 14, 2000, Change 1, May 25, 2001. As of March 3, 2017:
http://info.publicintelligence.net/CJCSM_3122.01_JOPES_Vol_1.pdf

———, *CJCS Guide to the Chairman's Readiness System*, Chairman of the Joint Chiefs of Staff Guide 3401D, Washington, D.C.: November 15, 2010. As of March 8, 2017:
http://www.jcs.mil/Portals/36/Documents/Library/Handbooks/g3401.pdf?ver=2016-02-05-175742-457

———, *Campaign Planning Procedures and Responsibilities*, Chairman of the Joint Chiefs Staff Manual 3130.01A,
Washington D.C.: November 25, 2014. As of March 3, 2017:
http://www.jcs.mil/Portals/36/Documents/Library/Manuals/m313001.pdf?ver=2016-02-05-175658-163

———, *Adaptive Planning and Execution Overview and Policy Framework*, Chairman of the Joint Chiefs of Staff Guide 3130, Washington, D.C.: May 29, 2015. As of March 3, 2017:
http://www.jcs.mil/Portals/36/Documents/Library/Handbooks/g3130.pdf?ver=2016-02-05-175741-677

Chapman, Dennis P., *Planning for Employment of the Reserve Components: Army Practice, Past and Present.* Arlington, Va.: The Institute of Land Warfare, September 2008.

Feickert, Andrew, and Lawrence Kapp, *Army Active Component (AC)/Reserve Component (RC) Force Mix: Considerations and Options for Congress*, Washington, D.C.: Congressional Research Service, December 5, 2014.

GAO—*See* U.S. Government Accountability Office.

General Algebraic Modeling System, Version 23.3, homepage, undated, accessed January 18, 2016. As of March 8, 2017:
http://gams.com/

Fontenot, Gregory, E.J. Degen, and David Tohn, *On Point: The United States Army in Operation Iraqi Freedom*, Fort Leavenworth, Kan.: Combat Studies Institute Press, 2004.

Hagel, Chuck, *Unit Cost and Readiness for Active and Reserve Components of the Armed Forces: Report to Congress*, Washington, D.C.: Office of the Secretary of Defense, December 20, 2013, pp. 5–15. As of March 2, 2017:
http://www.ngaus.org/sites/default/files/CAPE%20FINAL%20ACRCMixReport.pdf

Headquarters, First Army, "Mission," undated. As of March 6, 2017:
http://www.first.army.mil/content.aspx?ContentID=199

IBM, "CPLEX Optimizer: High-Performance Mathematical Programming Solver for Linear Programming, Mixed Integer Programming, and Quadratic Programming," undated. As of March 8, 2017:
http://www-01.ibm.com/software/commerce/optimization/cplex-optimizer

Joint Publication 5-0, *Joint Operational Planning,* Washington, D.C.: Joint Chiefs of Staff, August 11, 2011.

Klimas, Joshua, Richard E. Darilek, Caroline Baxter, James Dryden, Thomas F. Lippiatt, Laurie L. McDonald, J. Michael Polich, Jerry M. Sollinger, and Stephen Watts, *Assessing the Army's Active-Reserve Component Force Mix*, Santa Monica, Calif.: RAND Corporation, RR-417-1-A, 2014. As of February 14, 2017:
http://www.rand.org/pubs/research_reports/RR417-1.html

Lippiatt, Thomas F., and J. Michael Polich, *Reserve Component Unit Stability: Effects on Deployability and Training*, Santa Monica, Calif.: RAND Corporation, MG-954-OSD, 2010. As of February 14, 2017:
http://www.rand.org/pubs/monographs/MG954.html

Lippiatt, Thomas F., J Michael Polich, and Ronald E. Sortor, *Post-Mobilization and Training of Army Reserve Component Combat Units*, Santa Monica, Calif.: RAND Corporation, MR-124-A, 1992. As of February 14, 2017:
http://www.rand.org/pubs/monograph_reports/MR124.html

Maritime Administration, "Maritime Security Program (MSP)," undated-a. As of March 30, 2017:
https://www.marad.dot.gov/ships-and-shipping/strategic-sealift/maritime-security-program-msp/

———, "Voluntary Intermodal Sealift Agreement (VISA)," undated-b. As of March 8, 2017:
http://www.marad.dot.gov/ships-and-shipping/strategic-sealift/voluntary-intermodal-sealift-agreement-visa/

Marsh, John O., Jr., *Subject: Problems in Implementing the Army's CAPSTONE Program to Provide All Reserve Components with a Wartime Mission,* Washington, D.C., Secretary of the Army memorandum to the Army Inspector General, GAO/FPCD-82-59, September, 22, 1982. As of March 7, 2017:
http://www.gao.gov/assets/210/205773.pdf

National Commission on the Future of the Army, "Force Generation Subcommittee: Mobilization Force Generation Installation (MFGI)," information paper presented at meeting, Arlington, Va., November 5, 2015. As of March 3, 2017:
http://www.ncfa.ncr.gov/sites/default/files/MFGI%20Paper%205%20Nov%2015.pdf

National Commission on the Structure of the Air Force, *Report to the President and Congress of the United States*, Washington, D.C., January 30, 2014. As of March 3, 2017:
http://policy.defense.gov/Portals/11/Documents/hdasa/AFForceStructureCommissionReport01302014.pdf

Nemhauser, George L., and Laurence A. Wolsey, *Integer and Combinatorial Optimization*, New York: Wiley & Sons, 1988.

Pint, Ellen M., Matthew W. Lewis, Thomas F. Lippiatt, Philip Hall-Partyka, Jonathan P. Wong, and Tony Puharic, *Active Component Responsibility in Reserve Component Pre- and Postmobilization Training*, Santa Monica, Calif.: RAND Corporation, RR-738-A, 2015. As of February 14, 2017:
http://www.rand.org/pubs/research_reports/RR738.html

Public Law 102-90, National Defense Authorization Act for Fiscal Years 1992–1993, Title IV, Part A, Section 402, Assessment of the Structure and Mix of Active and Reserve Forces, 102nd Congress, December 5, 1991. As of March 3, 2017:
https://www.gpo.gov/fdsys/pkg/STATUTE-105/pdf/STATUTE-105-Pg1290.pdf

Public Law 108-375, Ronald W. Reagan National Defense Authorization Act for Fiscal Year 2005, Title V, Subtitle B, Section 513, Commission on the National Guard and Reserves, 108th Congress, October 28, 2004. As of March 3, 2017:
http://www.dod.gov/dodgc/olc/docs/PL108-375.pdf

Python Software Foundation, "The Python Language Reference," version 3.4.4, undated, last updated June 25, 2016. As of March 6, 2017:
https://docs.python.org/3.4/

———, "SimPy 3.0.10," undated, last updated September 1, 2016. As of March 6, 2017:
https://pypi.python.org/pypi/simpy

Rostker, Bernard, Charles Robert Roll, Marney Peet, Marygail Brauner, Harry J. Thie, Roger Allen Brown, Glenn A. Gotz, Steve Drezner, Bruce W. Don, Ken Watman, Michael G. Shanley, Fred L. Frostic, Colin O. Halvorson, Norman T. O'Meara, Jeanne M. Jarvaise, Robert Howe, David A. Shlapak, William Schwabe, Adele Palmer, James H. Bigelow, Joseph G. Bolten, Deena Dizengoff, Jennifer H. Kawata, Hugh G. Massey, Robert Petruschell, Craig Moore, Thomas F. Lippiatt, Ronald E. Sortor, J. Michael Polich, David W. Grissmer, Sheila Nataraj Kirby, and Richard Buddin, *Assessing the Structure and Mix of Future Active and Reserve Forces: Final Report to the Secretary of Defense*, Santa Monica, Calif.: RAND Corporation, MR-140-1-OSD, 1992. As of February 14, 2017:
http://www.rand.org/pubs/monograph_reports/MR140-1.html

Schlesinger, James, "Readiness of the Selected Reserves," Secretary of Defense memorandum, U.S. Department of Defense, August 23, 1973.

SDDC-TEA—*See* Surface Deployment and Distribution Command Transportation Engineering Agency.

Sortor, Ronald E., Army Active/Reserve Mix: Force Planning for Major Regional Contingencies, Santa Monica, Calif.: RAND Corporation, MR-545-A, 1995. As of February 14, 2017:
http://www.rand.org/pubs/monograph_reports/MR545.html

Surface Deployment and Distribution Command Transportation Engineering Agency, "Analysis of Mobility Platform (AMP)," U.S. Army, undated. As of March 30, 2017:
https://www.sddc.army.mil/sites/TEA/Functions/SystemsIntegration/ModelingAndSimulation/Pages/AMP.aspx

———, *Reserve Component Mobilization Process and Requirements for Installation Infrastructure*, April 2014.

Tucker, Michael S., statement to the National Commission on the Future of the Army, August 18, 2015, p. 6. As of March 3, 2017:
http://www.ncfa.ncr.gov/sites/default/files/NCFA%20First%20Army%20STMT%20FINAL_18_Aug_15.pdf

U.S. Code, Title 46, Chapter 24, Merchant Marine Act, 1920. As of March 30, 2017:
https://www.law.cornell.edu/uscode/html/uscode46a/usc_sup_05_46_10_24.html

U.S. Government Accountability Office, *Reserve Forces: Army Needs to Reevaluate its Approach to Training and Mobilizing Reserve Component Forces*, GAO-09-720, Washington D.C., 2009. As of March 8, 2017:
http://www.gao.gov/products/GAO-09-720

———, *Active and Reserve Unit Costs,* GAO-14-711R, Washington, D.C., July 31, 2014. As of March 8, 2017:
http://www.gao.gov/products/GAO-14-711R

Yuengert, Louis G., *How the Army Runs: A Senior Leader Reference Handbook 2015–2016*, Carlisle, Pa.: U.S. Army War College, 2015.